Innovation and Business Partnering in Japan, Europe and the United States

Collaborative alliances and business partnering offer a real alternative to mergers and acquisitions (M&A) as strategies for driving innovation and growth, particularly for small and medium-sized enterprises (SMEs) as they allow companies with limited resources to embark on the necessary research and development and find new markets. This book examines the role of business partnering as a pathway to innovation for SMEs, assessing the different models of business partnering in Japan, Europe and the United States.

It argues that SMEs will make or break the future of Europe and that a new model based on partnering for new ventures, high-technology and older non-tech companies is the only way forward. Using much material not examined before in English, it also reveals how important the SME sector is for Japan, and explains the different models the Japanese are pursuing to strengthen their SME sector. One of these models, innovation clustering, is akin to the American model, which has, through a series of partner relationships between private and public sectors at the local, national and federal level, provided SMEs with the financial and infrastructural support to drive the internal growth that prevents them from being absorbed by larger companies through M&A, which occur regularly in Europe.

Overall, this important book offers new perspectives on internal growth in SMEs, showing that high growth can be derived in any sector at all levels of technological advancement. It is essential reading for business people, economists, public administrators, managers and country specialists.

Ruth Taplin is Director of the Centre for Japanese and East Asian Studies, London, which won Exporter of the Year in Partnership in Trading/Pathfinder for the UK in 2000. She received her doctorate from the London School of Economics and is the author/editor of twelve books and over 200 articles. She has been the editor of the *Journal of Interdisciplinary Economics* for eleven years. Currently she is Research Fellow at Birkbeck College, University of London and the University of Leicester. She was appointed Visiting Professor at the School of International Business and Management, University of Warsaw, Poland in January 2005.

Routledge Studies in the Growth Economies of Asia

Innovation and Business Partnering in Japan, Europe and the United States

**Edited by
Ruth Taplin**

Routledge
Taylor & Francis Group

LONDON AND NEW YORK

First published 2007
by Routledge
2 Park Square, Milton Park, Abingdon, Oxon OX14 4RN

Simultaneously published in the USA and Canada
by Routledge
270 Madison Ave, New York, NY 10016

Routledge is an imprint of the Taylor & Francis Group, an informa business

Typeset in Times New Roman by
Florence Production Ltd, Stoodleigh, Devon
Printed and bound in Great Britain by
Biddles Ltd, King's Lynn

British Library Cataloguing in Publication Data
A catalogue record for this book is available from the British Library

Library of Congress Cataloging in Publication Data
  Innovation and business partnering in Japan, Europe, and the United States/
  edited by Ruth Taplin
      p. cm – (Routledge studies in the growth economies of Asia; 67)
      Includes bibliographical references and index.
    1. Strategic alliances (Business) – Japan.   2. Strategic alliances (Business)
  – Europe.   3. Strategic alliances (Business) – United States.   4. Small
  business – Technological innovations – Japan.   5. Small business –
  Technological innovations – Europe.   6. Small business – Technological
  innovations – United States.   7. Small business – Research – Japan.
  8. Small business – Research – Europe.   9. Small business – Research –
  United States.   I. Taplin, Ruth.   II. Series.
  HD69.S8I663   2006
  338.8'7 – dc22   2006016093

ISBN10: 0–415–40287–5 (hbk)
ISBN10: 0–203–96820–4 (ebk)

ISBN13: 978–0–415–40287–3 (hbk)
ISBN13: 978–0–203–96820–8 (ebk)

Contents

Illustrations

Figures

Table

Notes on contributors

Bernard Arogyaswamy is Madden Professor of Business and Director at the Madden Institute, New York. Prior to this he was Chair and Professor at Lemoyne College, Department of Business Administration, Syracuse, New York, and a consultant to SME business in Quality and International Business. He was a Fulbright Professor 2002–3 at the School of Management, University of Warsaw, Poland, and was awarded MBA Teacher of the Year. His first degree and industrial experience was in Madras, India.

Peter Davies is the Chief Executive of Pera International, a business that specialises in inspiring and enabling innovation in its client companies. It has recently expanded to twenty offices in twelve countries. Pera is now the hub of the leading European network to support manufacturing innovation and is the largest private-sector provider of international technology partnering services to British companies. Dr Davies has previously been a member of the British Cabinet's 'Think Tank', has been the Principal Advisor to the Government Chief Scientist, and in the early 1990s was the Director of what was then the largest joint industry–university technology centre in the UK. He is a member of a number of government advisory bodies. He has a first degree and Ph.D. in Mathemathics and Physics, and he was made a CBE in 2003 for services to business innovation.

Takuma Kiso has been Head of the Research Department in Public Policy at the Mizuho Research Institute Ltd since 2002. Prior to this he held positions as Senior Economist and Economist with the Fuji Research Institute Corp and with the Fuji Bank respectively. He holds a Master of Science degree in Financial Economics from the University of London and a Bachelor of Economics degree from the University of Tokyo, and he is a member of the Japan Association of Business Cycle Studies, the American Economic Association and the Japan–British Society. His recent books include: *Nihon Keizai no Shinro* (*The Future Course of the Japanese Economy*) (Chuokoron-shinsha, Inc., 2004, co-author). He has contributed articles to *Exploiting Patent Rights* *and a New Climate for Innovation in Japan* (Intellectual Property Institute, 2003), *Valuing Intellectual Property in Japan, Britain and the United States* (RoutledgeCurzon, 2004), *Shuukan Economist* (*Weekly Economist*) and *Insight Japan*.

Anthony Murphy is currently Director Strategy, Services and Operations Public Procurement Policy and Retail Business Relations DTI. This is after sixteen years in the private sector, much of which time he spent living overseas, was appointed to the Civil Service in 1992, in 1999 becoming Director of Copyright and an Executive Board member at the Patent Office, an Executive Agency of the Department of Trade and Industry (DTI). He was a member of the team that helped set up the DTI Innovation Group. In 2001, following the decision to merge the Patent Office's policy activities, he became Director of Intellectual Property and Innovation. From July 2003 to present he has been Director, Future of Europe, at the DTI. In this role he is responsible for DTI interests in European Union (EU) enlargement, treaty reform and the euro, as well as the 'Raising the Game in Europe' project, which aims to strengthen EU policy professionalism across the Department. He was closely involved in preparations for the UK Presidency of the EU in the second half of 2005.

Akio Nishizawa is Professor at the Graduate School of Economics and Management and Deputy Director of the New Industry Creation Hatchery Centre (NICHe) at Tohoku University in Sendai, Japan. He is also former Director of the Japanese venture capital (VC) organisation now known as JAFCO. Professor Nishizawa was also awarded the Bayh-Dole Award 2005, presented by Association of University Technology Managers (AUTM), for his contribution to the proliferation of US-style university technology transfer to Asian countries. His latest book concerns innovation clustering in Japan (in Japanese).

Alojzy Nowak is Dean of the School of Management, University of Warsaw, and is also Professor of Economics and Finance. He was educated in Poland, the US, the UK and Germany. He is also Director of the Centre for Europe, University of Warsaw; Chair in Finance, Kozminski School of Management, Warsaw; and Visiting Professor of International Business at a number of American and European universities, including the University of Illinois at Urbana-Champaign, the University of North Florida, the Free University of Berlin, and Bochum University, Germany. He is the author of over 120 books and articles published in Poland, the US, the UK, Belgium, Spain, Italy, Russia, Taiwan and Lithuania. Most of his research is concentrated on banking and finance, European integration, European monetary union, and transition in Central and Eastern Europe (CEE). He is a member of several national and international editorial boards and advisor to the Polish Prime Minister.

Ruth Taplin is Director of the Centre for Japanese and East Asian Studies, London, which won Exporter of the Year in Partnership in Trading/ Pathfinder for the UK in 2000. She received her doctorate from the London School of Economics and is the author/editor of twelve books and numerous articles. The most recent are: *Exploiting Patent Rights and a New Climate for Innovation in Japan* (Intellectual Property Institute, 2003); *Valuing Intellectual Property in Japan, Britain and the United States* (RoutledgeCurzon, 2004); *Risk Management and Innovation in Japan Britain and the United States Transition* (Routledge, 2005); and (with Masako Wakui) *Japanese Telecommunications: Market and Policy in Transition* (Routledge, 2006). Professor Taplin has been the editor of the *Journal of Interdisciplinary Economics* for eleven years. Currently she is a Research Fellow at Birkbeck College, University of London and the University of Leicester. She was appointed Visiting Professor at the School of International Business and Management, University of Warsaw, Poland in January 2005.

Terry Young has more than twenty years' experience in innovation management and technology transfer, most notably as Assistant Vice Chancellor for Technology Transfer for The Texas A&M University System (1991–2003), and Executive Director of its Technology Licensing Office. He is a Past President (2001) of the 3,500-member international AUTM (www.autm.net). In 2005 he resigned from the A&M System to become an entrepreneur, forming Terry Young Group, LLC. The focus of the new company is international development through the management of innovation and technology transfer. Since 2000, Young has organised or participated in more than forty-five international workshops on innovation management. He was named 'IPR Man of the Year for Nigeria' for 2004, and is one of only eight foreign members of the Czech National Academy of Engineering, elected in 2002 for his contributions to establishing a technology transfer infrastructure for that country. More recently, Young was one of only five private-sector members appointed to the US–Russian Innovation Council on High Technologies by the US Secretary of Commerce and confirmed by the White House.

Acknowledgements

I would like to thank Dr Peter Davies CBE for his assistance and support and for allowing access to his client base to make my chapter and this book a reality. I would also like to thank his staff and clients, who kindly shared their experiences, which afforded me greater insight into the requirements for SME companies in Europe to flourish. It reconfirmed the idea that those who practise often have the clearest understanding of the big picture.

In addition, I would like to thank Mr Michael Barrett OBE, CEO of the Great Britain Sasakawa Foundation (GBSF), Professor Peter Mathias CBE, President of GBSF, and the Earl of St Andrews, Chairman of GBSF, for their vision and continued support for the series of books I have been editing that explain different models for development of intellectual property (IP), innovation, risk management, financial services between Japan, Britain, continental Europe and the United States.

Lastly, thanks are in order to Mr Peter Sowden, Editor at Routledge for his valuable suggestions, Sir Digby Jones, Director General of the Confederation of British Industry (CBI), and Mr Anthony Murphy, then Director of the Future of Europe at the DTI, for their support of 'Can Europe make it? SME innovation partnering – the missing links', a seminar based on this report held on 5 October at Central Westminster Hall, London.

Ruth Taplin

Abbreviations

AI	artificial intelligence
ASEAN	Association of South East Asian Nations
ATP	Advanced Technology Program (US)
AUTM	Association of University Technology Managers
CAD	computer aided design
CBI	Confederation of British Industry
CBR	case based reasoning
CE	collective efficiency
CEE	Central and Eastern Europe
CIA	Central Intelligence Agency
CIP	Competitiveness and Innovation Framework Programme
CM	convergent milieu
CPA	certified public accountant
CRAFT	Cooperative Research Action for Technology
DRAM	dynamic random access memory
DTI	Department of Trade and Industry (UK)
EARTO	European Association of Research and Technology Organisations
EC	European Commission
EDA	Economic Development Administration (US)
EIS	European Innovation Scoreboard
EPC	European Patent Convention
EPSCoR	Experimental Program to Stimulate Competitive Research (US)
ERA	European research area
ERISA	Employee Retirement Income Security Act (US)
EU	European Union
FDI	foreign direct investment
FIRICS	Fault Identification Retrofication Intelligent Computer Vision System
FOMA	freedom of multimedia access
FP#	Framework Programme #
FTC	Federal Trade Commission (US)

GATT	General Agreement on Tariffs and Trade
GBSF	Great Britain Sasakawa Foundation
GDP	gross domestic product
GERD	government expenditure on research and development
GPS	global positioning system
GRP	gross regional product
GVA	gross value added
ICS	Industry Creation Section
ICT	information and communication technology
IIT	Indian Institutes of Technology
IMF	International Monetary Fund
IP	intellectual property
IPC	Innovation Policy Committee (Poland)
IPO	Initial Public Offering
IPR	intellectual property right
IRC	Innovation Relay Center
ISO	International Organization for Standardization
I/UCRCs	Industry/University Cooperative Research Centers
KBN	State Committee for Scientific Research (Poland)
LBO	leveraged buyout
LCD	liquid crystal display
LE	large enterprise
LSI	large-scale integrations
M&A	mergers and acquisitions
MCC	Microelectronics and Computer Technology Corporation
MDA	Metropolitan Development Assocation
MEI	Matsushita Electric Industrial Co., Ltd
MEP	Manufacturing Extension Partnership (US)
METI	Ministry of Economy, Trade, and Industry (Japan)
MEXT	Ministry of Education, Culture, Sports, Science and Technology (Japan)
MITI	Ministry of International Trade and Industry (Japan)
MRI	magnetic resonance imaging
NHS	National Health Service (UK)
NICHe	New Industry Creation Hatchery Center (Japan)
NIH	not-invented-here
NII	National Innovation Initiative (US)
NIST	National Institute for Science and Technology
NPD	new product development
NSF	National Science Foundation (US)
OECD	Organisation for Economic Co-operation and Development
OEM	original equipment manufacturer
PCB	printed circuit board

PD	power distance
PDP	plasma display panel
R&D	research and development
SBA	Small Business Administration (US)
SBDC	Small Business Development Center (US)
SBIR	Small Business Innovation Research (US)
SEMATECH	Semiconductor Manufacturing Technology initiative
SME	small and medium-sized enterprise
SoC	system-on-chip
SRAM	static random access memory
STREP	Specific Targeted Research Project
STTR	Small Business Technology Transfer Program (US)
TEG	Technology Enterprise Group
TFT	thin film transistor
TICP	Tohoku Intelligent Cosmos Plan
TLO	Technology Licensing Organization (US)
TLO Law	Law Promoting University-Industry Technology Transfer (Japan)
TNC	transnational corporation
TTA	Techno Arch Co. Ltd. (Japan)
UA	uncertainty avoidance
UNICE	Union des Industries de la Communauté Européenne
VC	venture capital
WIPO	World Intellectual Property Organisation

1 Introduction

Business innovation globally at a crossroads

Anthony Murphy

A central question we need to face in the UK and Europe is how well equipped we are to rise to the challenge of global innovation especially in relation to the rapidly developing innovators in Japan, the United States and Asia.

Europe has arrived at a crossroads. It may even be the kind of cross-roads Woody Allen had in mind when he said:

> More than any other time in history, mankind faces a crossroads. One path leads to despair and utter hopelessness. The other, to total extinction. Let us pray we have the wisdom to choose correctly.

We hope that the choice is not quite as desperate as that. But there is no doubt that the gradient Europe has to climb between now and 2010 is a truly daunting one. Whether we are inclined to view globalisation as a threat or an opportunity, there is no doubt that the climate in which policy is formed has changed dramatically since 2000, when European leaders met in Lisbon and set themselves the ambitious and possibly even deluded target of making Europe the most dynamic and competitive knowledge-based economy in the world by 2010. Halfway down the road, it is fair to say that the Lisbon bubble has been punctured at several points.

The Asian challenge – Japan, China, India

In the near future the world will be in the grip of the claws of the Chinese Dragon and the Indian Tiger. Shorter product development lead-times, new technologies, new ways of doing business and, crucially, new ways of *thinking* that owe little to the past mean that more and more companies are sourcing the inputs they need wherever costs are lowest. But although lower labour and other costs are clearly one element in the challenge Europe faces from countries such as China and India, it is in the area of innovation that the Asian economies are arguably having the biggest impact. At a time in our history when the stock of scientific knowledge is doubling every five to seven years, and when 90 per cent of all the scientists who ever lived are alive today, India produces more science graduates each year than the whole of the European Union (EU). In 2004 China and India between them produced 125,000 *computer* science graduates; the comparable figure for the UK was 5,000. China has tripled its spending on research and development (R&D) over the last five years; and India plans to quintuple the size of its biotechnology sector over the next five years.

This is independent of the surge in innovation through the past five years of the intellectual property (IP) drive taking place in Japan. In 2005 Japan was second only to the United States, producing 25,145 and 45,111 patents respectively. The European Patent Convention (EPC) member states produced the most only when calculated together for all states at 46,446 (these are World Intellectual Property Organisation (WIPO) estimates). However, the East Asian countries showed the most impressive growth of all, with Japan, South Korea and China alone accounting for 24 per cent of all applications, and with Japan registering the most, showing an annual rise of 212 per cent.[1]

The US remains a formidable economy

It is important not to be myopic about the Asian challenge. Let's not forget that, even if China does indeed overtake the US and become the world's largest economy in 2039, as Goldman Sachs has predicted, the US remains a formidable competitor and partner, outperforming the EU in nine out of twelve innovation indicators. The European Innovation Scoreboard (EIS) identifies the following key factors in this widening gap:

* patents;
* size of the working population with tertiary education;
* R&D expenditure;
* early stage venture capital (VC).

Never a country to stand still, the US launched its National Innovation Initiative (NII) in 2004, and has implemented a 4.6 per cent increase in the federal science budget so that it now amounts to $100 billion per annum (including defence-related research, which has traditionally been the source of the most innovative research in the US).

How should Europe respond?

Against this background the temptation for Europe to don sackcloth and ashes and sink into a swamp of self-doubt and self-loathing is a strong one; as Gunter Grass said, 'building of castles has always been our special joy, / To raise the rampart, excavate the moat'. But this is not a time for invertebrate behaviour. Knowledge – the alloy of innovation, education and research – is central to Europe's future economic success and, by extension, to the fulfilment of Europe's social and environmental ambitions. Several nettles have to be grasped here: getting Europe's spending on R&D closer to the 3 per cent target set in Barcelona in 2002; and incentivising business to spend more on R&D, through measures such as the UK's tax credits scheme, but also and crucially through much-needed reform of our IP regimes, both at the national and European level.

So innovation is the key to survival. And innovation needs to be broadly defined, encompassing not just R&D but also new processes and business models, new ways of managing knowledge and new approaches to forging partnerships. Research in the UK suggests that R&D currently accounts for only 40 per cent of expenditure on innovation, which is increasingly invested instead in design, skills, brand identity and marketing.

Britain has taken significant steps to support innovation, not least by investing heavily in its science base, notching up £1.3 billion through R&D tax credits, and using the government's massive spending power to stimulate and support innovation through the government procurement process. Following a report by Jean-Louis Beffa, Chairman of St-Gobain, of the Industrial Innovation Agency, France saw the creation and launch over the summer of 2005 of 'competitiveness clusters'; Michael Porterism is clearly alive and well in cities such as Paris and Toulouse, Bordeaux and Lyon, Grenoble and Marseille.

European countries do, of course, have different starting-points, which militates against a 'one size fits all' approach to the innovation challenge. The European Commission's (EC) Summary Innovation Index shows, for example, that the UK scores well on higher-level skills and the number of science graduates it produces; but it lags behind Germany on the level of expenditure on R&D.

The R&D Framework Programme

It is at the *collective* European level that the major challenges lie. And many of those challenges coalesce around the R&D Framework Programme (FP) – Europe's third largest spending programme and the principal mechanism for encouraging and supporting R&D. One of the successes of the UK Presidency of the EU in the second half of 2005 was to secure a partial general approach on FP7, in other words, Council agreement on the Programme's scope and governance. Now that the Financial Perspectives for 2007–13 have been agreed, the way is open for converting the partial agreement into a complete package in 2006.

Clearly FP7 is about more than just money. If it is to succeed, it must be structured and delivered in ways that are transparent, easy to understand and sensitive to the fact that business people have better things to do with their time than wade through a glutinous morass of bureaucracy. We hear a lot nowadays about simplification and the wind of change (more of a light breeze, perhaps) that is blowing through the Commission, or at least those parts of it obedient to the will of Vice-President Verheugen. But that wind needs to infiltrate the Framework Programme too. Research by the UK's Department of Trade and Industry (DTI) shows that four-fifths of British businesses that took part in FP4 did not then go on to participate in FP5. Private-sector R&D will always be more volatile than its public sector counterpart. But flagging business enthusiasm for the Framework Programme must have something to do with business aversion to red tape and labyrinthine processes, too. However, the oversubscription by small and medium-sized enterprises (SMEs), in particular, for the FP6 programme through innovation facilitator companies such as Pera has shown a great willingness on the part of the companies to innovate if another organisation completes the onerous paperwork. SME companies, predominantly non-high-tech and established companies comprise the bulk of potentially innovative, profit-making enterprises in the EU. A very recent decision by the EU in February 2006 has extended the budget so that many more non-high-tech established companies can participate in FP7.

Better regulation

Innovative individuals and businesses can only realise their full potential within a regulatory framework that is genuinely supportive, intelligently designed and easy to understand. Regulation needs to be squarely based on consultation, high-quality evidence, a proper assessment of risks, and a readiness to consider alternative approaches that achieve the same objective, with both costing and drafting carried out with an eye to Europe's international competitiveness.

Better regulation was a major cross-cutting policy theme during the UK Presidency in 2005, and one that attracted broad support around the Council table. But when I began my Civil Service career in 1992 *de*regulation (as it was then called, with little thought for the discordant effect this often caused) was widely seen in continental Europe as a peculiarly British vice, akin to warm beer and cold showers. But in a remarkably short time we have witnessed the re-branding of deregulation as better regulation (via 'Simpler Legislation in the Internal Market' and other initiatives) and its emergence as a central element in the European economic reform process. Influential advocacy has helped: from the EC, which has declared regulatory reform as a key element in achieving the Lisbon goals; and from the International Monetary Fund (IMF), which believes improvements in the EU regulatory framework could deliver as much as a 7-per-cent increase in gross domestic product (GDP) and a 3-per-cent increase in productivity in the longer term.

With regard to the subject of better regulation, a sentence from a book written by the historian John Julius Norwich in 1977 encapsulates this idea:

> They might feel irritated, from time to time, by the petty regulations and restrictions through which the state sought to interfere with so many aspects of their daily life; but if this was the price of living in the richest, safest, best ordered and most beautiful city in the civilised world they were prepared to pay it.

The city referred to here is Venice in 1400, a city where the specification of merchant ships was defined in minute detail so that Venetian ships were faster, stronger, more competitive, more likely to reach their destinations with their cargoes intact than those of their trading rivals; where consumer confidence was maintained by officials whose job it was to ensure that customers were never overcharged and wine never watered down – in other words, innovation and regulation in perfect harmony.

There are encouraging signs that this benign virus is spreading, not least through the veins and arteries of the Framework Programme. The Commission has conceded that the Programme needs to be more user-friendly, and it is looking at various ways of improving the situation, including new management and delivery mechanisms and clearer guidance. FP7 has a simpler structure than its predecessors, built around the four themes of cooperation, ideas, people and capacities (including under this heading some welcome activities aimed at ensuring greater participation by SMEs). The challenge is to replicate this simplicity not just in the design of FP7 but in its delivery as well.

And we need to apply this same discipline to the Competitiveness and Innovation Framework Programme (CIP). On the face of it, the idea of bringing together a hotchpotch of existing support programmes[2] worth more than 4.2 billion euros into a single, unified programme is a good one. Besides presenting business with, in effect, a supermarket rather than an archipelago of small corner shops, the CIP should also help to avoid overlap and duplication across a broad tract of European policy, including the Framework Programme and the Structural and Cohesion Funds.

The focus within the CIP on entrepreneurship and innovation, in particular among SMEs, is greatly welcome. It will nonetheless be important, as with FP7, to make sure that sufficient transparency and flexibility is build into the three sub-programmes (Entrepreneurship and Innovation, ICT Policy Support, and Intelligent Energy-Europe, each with its own management committee) if synergies are to be identified and exploited.

Open innovation

Both FP7 and the CIP, and indeed all the policy interventions made over the life of the next budget and beyond, need to acknowledge that the nature of innovation is changing. Henry Chesbrough has written about this evolution in his book *Open Innovation*. Open innovation is about speed and agility. It's about grabbing the best ideas and the best people wherever you can find them. It's about seeing innovation not just as new products and services, but new processes, new business models, new ways of working. It's about stripping IP of its rather dusty image (after all, no one likes intellectuals, and all property is theft) and knowing to the last penny the value of the IP you own. It's about networking and interaction and partnership, multiple sourcing of ideas and the free flow of skills and experience. No man and no business is an island in the open innovation world.

The closed innovation model remains tenuously relevant in some industries, but its broader ascendancy is at an end. The open innovation model is better placed to cope with the often unpredictable pace and direction of technology, and the volatility and growing sophistication of new markets and consumer behaviours.

The role of information and communication technology

The sea-lanes that pull these markets and behaviours together are digital networks. A report commissioned by the DTI and published last year (*i2010 – responding to the challenge*) argues that policies that inhibit the productive use of information and communication technology (ICT) played a major part in Europe's relative economic decline over the past decade. Over the five years from 1996 to 2000, ICT contributed almost three times as much to labour productivity in the US than it did in the EU. The proposed solution involves not just greater investment in ICT capital and skills but also the shaking up of the ways businesses organise themselves and their activities – what Joseph Schumpeter has famously called 'creative destruction'. And for national governments and the EC the challenge is to build regulatory frameworks that allow all this to flourish – not least a state aid framework that is tolerant of policy interventions that support the creation and growth of innovative start-ups, increase the stock of risk capital, help SMEs to recruit the people they need and to buy the services they need, such as training and IP consultancy and incubation support.

The deployment and exploitation of ICT is a critical lever for enabling innovation, productivity improvement and economic growth. But both in the UK and in the wider EU, there are worrying signs that opportunities to exploit the potential of ICT have not been taken up as enthusiastically as they have in the US. There is an urgent need for governments to:

- develop a shared understanding across governments, business and other key stakeholders, of the ways in which ICT impacts on economic performance;

- identify the barriers that exist to the successful harvesting of the economic and other benefits flowing from the take up of ICT technologies;
- assess which barriers can be removed or reduced through partnership between industry and government, the aim being a regulatory framework that promotes market development and encourages investment in technology, innovation and skills;
- identify, with business and other stakeholders, what more they need to do to improve ICT-related productivity and align the actions of business and government to achieve the right results, including measures targeted on raising confidence in the use of ICT;
- ensure 'Joined Up Government' across the range of government activities affecting both the ICT industry itself and the adoption of ICT technologies, taking into account the role of government as a driver of ICT-enabled transformation through focused infrastructure investment and procurement.

Conclusion

We are at a crossroads. What road has Europe travelled in relation to Japan, China and the United States? Europeans have always had a flair for innovation. And by innovation I mean not just the successful exploitation of new ideas but also the translation of those ideas into tangible improvements in the quality of people's lives. Think of how the Portuguese blithely pillaged other people's ideas – maps, navigation tools, the lateen sail – to push the tentacles of European trade deep into Asia in the sixteenth century. Remember how in the early nineteenth century Britain ruthlessly exploited the comparative advantages it enjoyed over continental European rivals – internal security, international trading networks, high levels of legal protection for property, a sophisticated banking system and a seething, brimming talent pool of scientists and engineers – and how these assets were used to build a globally competitive economy out of non-British inventions: the Jacquard loom, the Leblanc soda process, Berthollet's chlorine-bleaching technique and Koechlin's dyes: plenty of 'open innovation' there. And though we worry about links between the university research base and business, how easily we forget experiments in collaboration such as the Solvay Conference in Belgium in 1911, which brought leading industrialists together with Albert Einstein, Marie Curie, Max Planck and Ernest Rutherford.

Now compare this European inheritance with another great civilisation – one that invented the magnetic compass but used it for interior design rather than oceanic travel and trade; that cast intricate bronze statues of extraordinary beauty but never developed high-precision engineering; that invented paper and printing but ruthlessly suppressed knowledge transfer; that burned coal but never made a steam engine; that had a single market, a single currency, an efficient civil service recruited by competitive examination – all contained within a wall 1,400 miles long: China during the Ming Dynasty, trapped in what David Landes memorably called 'the magnificent dead end'. *Civilisations do die*, and they die because they fail to innovate, because they fail to compete, because they believe they are better and more blessed than their rivals. We cannot make the same mistake. Japan – still the second largest economy in the world after a decade of stagnation – has realised, just as Europe is beginning to, that intensive energy and resources have to be invested in value-added innovative products and services and that business partnering is the sure way to overcome excessive research costs, pool expertise and talent and share client/supplier databases.

Europe is well equipped by its own history to rise to this challenge: did Europe not give the world the concept of Enlightenment, what Kant defined as 'liberation from self-imposed tutelage'? Where better than here to emancipate ourselves from old ways of thinking, old ways of making and implementing policy?

Notes

1   This information is from the public information website of WIPO, Press Release 436, Geneva, 3 February 2006.
2   Multi-annual programme for Enterprise and Entrepreneurship; Intelligent Energy-Europe; promotion and demonstration of environmental technologies under the Life programme: Modinis, e-content and e-TEN programmes; certain innovation-related elements of the R&D Framework Programme.

2 Can Europe make it?

SME innovation business partnering – the missing links[1]

Ruth Taplin

The bulk of the material in this chapter was taken from the report researched and written by Ruth Taplin, 'Can Europe make it; SME innovation partnering – the missing links', presented to a distinguished audience at Central Hall Westminster on 5 October 2005. Sir Digby Jones, Director General of the Confederation of British Industry (CBI), was the keynote speaker.

Many large companies in Europe are explicitly moving to a new model of corporate innovation, proactively building a global network of innovation partners and setting up cost-sharing innovation consortia moving away from growth gained through mergers and acquisitions (M&A). Growth through business partnering is becoming a viable alternative to M&A and a process that can reverse Europe's tendency to continue failing to mobilise its SME innovation and growth potential. A number of well-defined reasons for this failure were brought to the fore through a cross-section of interviews of SME companies[2] in the UK, Scandinavia and Eastern Europe. These companies, which are successful, range from low to high technology. Contrary to the myth that only high-tech companies can be high growth, the study on which this chapter is based showed that well-established low- to medium-tech companies rank among the highest growth companies. The companies interviewed showed that:

1   Not enough research is carried out by SME companies in Europe.
2   There is too little emphasis on market driven and interdisciplinary research.
3   Insufficient investment is being made in knowledge-intensive business development among the SME sectors.
4   Not enough young people are being encouraged to take on research careers and those that are do not have enough knowledge of how to innovate and successfully encourage SME companies to do the same in their future roles as business leaders.
5   Despite the impetus given by business pressures and the European FP6 and the up and running FP7, such programmes are not sufficiently focused on the objectives of developing new products with funding

and advice back-up at the point of commercialisation and the protection of IP.

6   A climate of failure being absolute is being promulgated, not allowing and encouraging those who have failed to get back on their feet and try again and again to succeed without being for ever branded as failures.

7   Although innovation by consortia of SMEs on an increasingly wider scale has been happening in Europe, it has not been sufficiently recognised or heralded. More cross-national consortia R&D needs to be undertaken that involves the new member states. The consortium innovation that does exist is not given enough recognition, and this includes the essential role of external innovation facilitators, which – this report demonstrates – are crucial to both SME and mid-sized company success.

8   When SME companies become mid-sized companies they are branded successful and no longer in need of assistance, causing them to fail because they fall between two stools.

Business partnering innovation – the missing link

More effective involvement of SME companies in the FP7 is essential if Europe is to raise R&D investment and reach the objectives of: (1) 3 per cent of GDP (Barcelona objectives); and (2) the globally most competitive knowledge-based economy (Lisbon objective). Companies that were interviewed stated very clearly what it was that FP6 offered (explained in the case studies at the end of this chapter) that assisted them to move forward as European SME companies. The FP7 budget has also been expanded to include more SME companies and those with high growth potential from the low- to medium-tech sector.

The most often cited assistance that companies mentioned in the interviews was related to point 7 above, that of business partnering innovation and external facilitators, with partnering taking place inside or outside their normal supply chains. Just as the cost of R&D for high-tech companies is beyond the capabilities of just one company, so, too, is the task for the vast majority of medium- and low-technology SME companies (97 per cent of small firms) with limited resources progressing towards higher-value products and being more knowledge intensive without pooling their resources with other companies. The international dimension is also vital, as no company or nation has a monopoly on good ideas. In fact, the wider the pool to search for those with innovative ideas, the more likely such originality will be found. In other words, international innovation partnerships give rise to new international business opportunities. As research knowledge is very specialist, it follows that SME companies will need to look beyond their national boundaries to link up with the best companies and technology service providers to contribute. The best manner in which to do this – as shown by the report's case studies – is to undertake a project together with an innovation facilitator or 'innovation coach' who already has a network of research contacts to chose from including companies, institutes and universities or even other *innovation facilitators*.

Companies interviewed also realised the richness and abundance of available resources they were sitting on in terms of existing suppliers or other associates who held the research information or practical expertise that was lacking in a new project or undertaking. The missing links, therefore, are all the existing knowledge-rich contacts that are available with a little strategic thinking and planning and assistance from the user friendly innovation facilitator and/or European research database. Pooling existing expertise and contacts keeps the costs low and in the case of the European Framework Programme, can also attract cash to match the company effort to fund the hiring of world-class technology developers to create the solutions they need. Innovation facilitators who can help small firms create the new product ideas and facilitate them through the 'consortium innovation' process are seen as being the next major tool for industrial growth and gross value added (GVA) improvements to industry – so much so, in fact, that several regions in the UK (East Midlands, East of England and Yorkshire and The Humber) have launched programmes to subsidise the provision of such support, as have several of the UK's European neighbours who are launching national schemes, including Spain, Ireland and Norway. UK manufacturing, which has been waning in the last few years, has in fact been assisted by its European neighbours who are both investing and buying more of the UK's manufactured goods.

Innovative consortia do not just have to work on one level, such as just sharing basic technical knowledge and skills. Companies interviewed showed that partnering to access a market that one company knows better than the other is useful, as is having an innovation facilitator share their wealth of experience in creating new products and services to improve on an innovative idea inside a company. Innovative business partnering also bypasses excessive bureaucracy, fosters cooperative partnering where all are winners and in this manner challenges the blame, failure and envy culture so entrenched in Europe and particularly the UK. Another point is that large-scale projects such as the Airbus A380 were the result of business partnering innovation, with different parts of the airplane produced by many different EU countries – for example, the wings were made in the UK and designed in France – but all the publicity surrounding it emphasised national triumphs instead. The essential message of the successful business partnering of many European countries and companies was completely lost.

In April 2005 Business Britain awarded Pera, one of the world's leading centres of innovation know-how, the Regional Business Support Service of the Year accolade. Their CEO, Peter Davies, describes some of the features that make external facilitators invaluable to innovation inside SME companies, paraphrased in the next few paragraphs.

Technology transfer in itself does not allow existing companies to absorb and implement effectively the new technological knowledge created regionally, nationally or internationally. Senior managers need to be given the tools to aid them to conceive, visualise and develop ideas for new innovative products. When this sort of support is accessed, impacts on the firms are frequently as high as doubling or even trebling their sales in new and higher-value markets. The majority of the fastest-growing technology-based companies are in the established sectors in which senior management have been able to plan confidently through an innovation programme.

Innovation facilitators such as Pera are private-sector companies that work in the public interest. They have battle-hardened capabilities and deep experience built and developed with international clients over many years that can be imparted directly to small and medium-sized local firms. The public interest aspect arises because profits are not distributed but are invested back into making their services better.

Service and advice based simply on opinion, however, are not of any great use as too many business decisions are clouded by competitors' actions, existing technologies, the latest popular trends or uncertainties over markets, rather than good, sound fact. Pera, as an alternative, does not base its service on opinion but evidence in the form of the knowledge centre staffed with business and technology analysts who find the facts and help to facilitate innovative action with a passion and commitment.

Missing links at different levels

Another missing link, particularly in the Framework Programme, is the lack of collaborative funding and support after the idea has been taken to the point of pre-commercialisation. In other words, the product is technically developed through the aid of an innovation facilitator on a consortium basis, but the commercialisation of the product is left to the SME company at one of the most difficult phases of product development: how to present the product to the market (e.g. how to present it to potential clients when it has not yet been tried and tested on the actual market). The product must be refined, presented, valued, patented (if applicable) and eventually licensed, when operating in a cross-border environment. The author notes that some of the companies suggested a process akin to a commercialisation and marketing valuation pool. Companies could try and test each other's products. Valuing the worth of a product, whether it be for a brand or copyright, is an art not a science, and the pooling of experiences in this field can only enhance that process.

The growth of e-business

Another missing link is the neglect of e-business. ICT and e-business used to be cited as important areas to facilitate innovation. E-business can be a real aid to SME companies that do not possess the resources to travel to visit clients as often as they may wish to do. E-business can also be used for networking, allowing greater access to untapped resources that other companies hold for consortia use and patent pools.

Estonia is an example of an entire country that has become an e-republic. For the study that this chapter is based on, one Estonian software company was interviewed and also a senior expert who runs the business and development section of Enterprise Estonia, which will be featured in case studies later in this chapter. The software company was part of an innovation consortium working with Pera under the FP6 and including companies from Britain, Denmark, Norway, Sweden and Estonia. Having visited Estonia it was not surprising that an Estonian company was involved in a consortium of high-tech security scanning, and the official at Enterprise Estonia proved to be just as dynamic as those I met in the company.

Since Estonians are reserved and not great self-publicists, it was at first easy to underestimate the drive and curiosity that pervades this nation state. Never content with the ways things are, the Estonians are always looking for new solutions and ways of developing ideas and products. Processes not moving quickly enough lead to frustration and a drive to move more quickly.

The roots of the collective expertise in software programming, cybernetics and artificial intelligence (AI) lie with the Soviet years of rule, when, in order to prevent any incipient movements towards independence, Estonia was made a centre for AI, and much of the software programming and development for the Soviet space programme was conducted there. In 1991, when Estonia became independent, it pursued a 'Tiger Leap' programme that integrated computer literacy and information technology into all national programmes and institutions. With access to the internet relatively inexpensive by that time and with no huge infrastructural investments to be made, some 52 per cent of Estonians now use the internet regularly. Almost everything, including government cabinet meetings, is efficient and rapid without paper. Two years ago an optional smart ID card that is making documents and money obsolete was introduced and is being embraced by the majority of Estonians.

Estonia is reminiscent of Singapore during its drive to become an e-city nation state. The small population of 1.4 million coupled with a good grasp of English by the young and the majority being connected to the internet has led to a growth rate of 6 per cent of GDP while the rest of Europe languishes at under 2 per cent. As Singapore demonstrated, it is much easier for a young industrious city state to become a successful e-republic, making decision-making a much quicker and efficient process. Singapore has become a model for parts of Mainland China such as Pudong in Shanghai. Estonia can be an example to the rest of Europe, connecting to other parts of Europe and serving to promote innovation consortia, entrepreneurship and high-tech proficiency, especially in SME companies.

A comparison with advanced nations

A draft report by a thirteen member group appointed by the EU and led by Wim Kok, the Dutch prime minister, has stated that the Lisbon agenda, which had as its target to make the European economy more flexible and entrepreneurial by 2010, is failing and that there is even a threat to the whole existence of the EU. The draft report warned that such economic failure could be devastating to Europe, undermining its civilisation, and that if Europe cannot adapt, the ageing working population will not be able to maintain pension levels, causing any potential economic growth to stagnate. The enlargement of the EU will also cause great difficulties if overall European economic performance does not improve. Worse still, despite the forward looking intentions of the Barcelona and Lisbon accords, Europe is continuing to lose ground to both the United States and Asia.

In relation to research, innovation and competitiveness, Europe's greatest weakness compared with the US is its SMEs. While both the total public expenditure on R&D and the research activity of large companies in Europe and the US are broadly comparable, SMEs in Europe undertake between seven and eight times fewer research activities than their American counterparts. An analysis by an EC staff economist shows that this difference between SME sectors substantially explains the R&D gap between Europe and the US.

Another missing link brought out from interviewing the companies in the study this chapter is based on is that SME companies are lacking step-by-step methodological advice about how to identify and absorb technological solutions into firms once an innovative product idea has been created to produce the 'market-pull' for the technology transfer process. One potential reason for the high level of research intensity in US small companies is that they may be provided with better support, not just through the innovation process but in the accessing and profitable integration of the results of research into the product ideas generated by the innovators inside the firms.

A recent manual published in America includes an explanation of the critical roles of: the inventor, licensing associate and invention advocate; laws, policy and procedures; understanding and managing the process; the IP strategy; the marketing strategy for inventions; negotiating the terms; and monitoring compliance and maintaining the relationship.

Even in Japan, which has seen a decade of economic underachievement, deregulation is occurring, rather than the increasing and stifling red tape and bureaucracy that Europe is now producing to the consternation of its business community where any gains from incremental regulation are being outweighed by the costs. The Japanese government is allowing businesses to work more directly with inventors from both the university and the institute sectors, encouraging them to start their own Technology Licensing Organizations (TLOs) and funding intra-departmental university offices that offer in-depth advice and practical assistance with developing innovation ideas and products at all levels. This process is at an early stage, and it seeks to allow inventors more freedom while working within the constraints of an institutional base, as the Japanese system is not organised to deal with individuals. This does not mean however that, as has been proven in Europe, groups of external innovation facilitators such as Pera or the Fraunhofer Institute may not be deemed in the future as institutions that can deal with other institutions to promote company partnering.

Europe's comparative economic weakness is all the more acute when one considers that SMEs account for 65 per cent of European GDP but only 45 per cent in the US. Another striking difference is that 75 per cent of large firms founded since 1980 in the United States have grown from small beginnings. By contrast, 80 per cent of similarly aged large firms in Europe are the result of M&A. Europe needs more innovative SMEs that grow into mid-sized companies that are successful in themselves.

The necessity of European funding for cross-border R&D

Most SMEs begin trading domestically. Many SMEs in Europe have international potential but too few of them have opportunities to gain effective connectivity for their product and service offerings into European and world markets. In a rapidly globalising business environment, this represents a strategic weakness for Europe. Participation in European projects to develop new technology to enable innovative products and services gives SMEs an important opportunity to learn about international markets and build the transnational connectivity to supply into them.

Most regional and national programmes that fund SME R&D in individual member states do not support the research activities or business partnering of their SMEs beyond their national frontiers. Pan-European programmes with sufficient resources, such as the Framework Programme, are essential to fill this gap. At the same time, other transnational initiatives such as Eureka have so far failed to fill the gap for SMEs due to inadequate budgets and poor coordination between member states in Europe. Hence it would be a strategic political failure, entirely contrary to the Barcelona and Lisbon objectives, not to provide adequate funding in support of the internationalisation of SMEs with European and global growth potential through Framework Programmes.

Refocus efforts to better sustain competitiveness and promote growth

A dynamic SME sector is essential to a modern economy, and governments have a key role to play in setting the right conditions to ensure that SMEs receive appropriate assistance for their knowledge development and business growth. Europe's weakness compared with the US lies not necessarily in the volume of public support for SME research but in its targeting. The current Framework Programme focuses 80 per cent of its target of 2.25 billion euros for SME participation on *high-tech SMEs* that are capable of conducting research at a high level of excellence but that often have little or no manufacturing capability of their own. However, these firms represent just 3 per cent of the total SME population and frequently are more research-driven than market-oriented or are locked into the value chains of large enterprises (LEs). SMEs with a focus on their own products and core business growth were largely deterred from FP6 participation by the new instruments' emphasis on longer-term and large-scale research. In the large integrated projects they are disadvantaged to such an extent that nearly one-third of British SMEs considers the cost-benefit ratio of participation as negative. Therefore, continuing to focus the vast majority of European research funding on this tiny population of high-tech, high-growth gazelles that are already research-intensive but that often lack high-growth potential is very unlikely to help progress towards either the Barcelona or Lisbon objectives. By contrast, there is great scope, still much under-exploited, for the Framework Programme to stimulate knowledge and business growth among the much larger community of existing *medium-tech SMEs* (30 per cent of the population) able to conceive innovative, technology-enabled products and services. While these firms do already participate enthusiastically in FP6, through a very popular programme called Cooperative Research Action for Technology (CRAFT) and to some degree through Specific Targeted Research Projects (STREPs), about 90 per cent of applicants fail to get funding as the scheme is more than ten times oversubscribed. This will change to a degree, given the decision in the early part of 2006 to provide more funding for SME gazelles irrespective of length of establishment or level (e.g. low, medium or high) of technological innovation.

Interviews with the main half-dozen European trade associations, including Union des Industries de la Communauté Européenne (UNICE), the European version of the UK's CBI, revealed a passionately held collective view that SME funding from Europe should be provided more on the basis of the quality of the innovation and exploitation potential of the ideas than on the scientific excellence of the research proposed. Dissemination and demonstration should be an integral part of the funding regime in order to maximise the potential for research results to become integrated widely into products and services able to reach world markets more often, more rapidly and more extensively than now. Hence a more even balance (say 50:50) between these two forms of SME support would produce a step change in the effectiveness of Europe's investment in SME research. The levels of funding for high-tech SMEs participating in the Framework Programme's main project types (integrated projects and STREPs) need to be reined back and balanced with a much greater level of funding for medium-technology SMEs that have much greater growth potential and use the CRAFT scheme.

The role of the EU

A number of vexing questions remain that are in need of urgent answers. Can the EU Framework Programme take account of SME requirements as outlined above and at different levels? What role should local/national governments play if any, to facilitate success for SME companies and a growth rate of at least 3 per cent across Europe? What methodology used successfully in the US could be followed in Europe? This is discussed in Chapter 5.

The EU will make available over 73 billion euros between 2007 and 2013 to facilitate the reality of the targets set by the Lisbon accords. Given the answers from the companies, trade associations and innovation facilitators interviewed it becomes amply clear that if Europe is to achieve the impact it needs from this investment, national, regional and European government agencies need to recognise that:

1 A clear link needs to be made between R&D and entrepreneurship.
2 Innovation should be market led, and not driven by the generation of new scientific and technological knowledge, as is currently the case. Regional and national support mechanisms need to be put in place to help firms with the new product ideas that will create the demand for new scientific and technological knowledge.
3 While keeping an eye on the market, opportunities and risks need to be viewed within a flexible context and with vision – in many cases demanding a behavioural change in attitude towards the relative importance of innovation within many SME companies in the UK.
4 Partnering is essential to enable SME companies to new levels of strategic thinking research/knowledge intensity and economic growth.
5 Partnering needs to be used to open new routes to markets and suppliers.
6 Innovation facilitators operating within government-supported programmes at a national and/or a regional level are likely to play a key role in achieving the changes in SME company behaviour, value added and sustainability. The use of innovation facilitators in a few pioneering regions should be studied more closely in order to identify the key skills that need to be imparted to help SME companies

to generate and bring out from company members vision and passion, and to offer direction while pointing out opportunities and helping to arrange partnering.

7   When SME companies are in the market they should be left alone, but when they fail some government back-up should exist as SME companies do not have the resources to take high risks to re-establish themselves. To provide a global example, the Development Bank of Japan has re-emerged as a new driving force in Japan, using a bank-rupt or failing company's IP as leverage to assist it with re-starting at medium or high levels of risk. The programme has been very successful and the European Construction and Redevelopment Bank could serve a similar role. This does not mean, however, that the emphasis for assistance should rest with failed companies – on the contrary, the high-growth potential companies need to be encouraged at every technological level.[3]

8   Simply throwing money at SME companies is counterproductive. Resources given out need to be carefully targeted and results need to be followed up meticulously.

9   SME companies need to be encouraged to spend within their budgets to reduce risk and create extra value. These are skills that need to be imparted by experienced innovation practitioners at the level of commercialisation as understanding the value of a product/IP is essential to market success.

Finally the establishment and management of a business innovation consortium has some very different features to normal supply chain management – there are pitfalls for a company inexperienced in this area. The chances of a successful expansion depend on a clear strategy at the outset but also on:

•   having a methodical and thorough risk analysis of the total route to supply the product or service in volume in a new market and with potentially new suppliers;
•   having the best possible partner-search process and professional assistance in business partnership building;
•   working with a facilitating organisation that can solve many of the induction problems in an unfamiliar territory that could be pan-European or global.

The domestic market is limited and creating regional hubs of excellence

Partly driven by the business pressures referred to above and partly stimulated by programmes administered by the EC to improve competitiveness, thousands of SME companies are now expanding their business outside their domestic market. With its sites in six member states, Pera has been a leader in making this happen and has established more pan-European industrial consortia for product and process innovation than any other organisation, and the pace is increasing. Staying within a local, domestic market limits the scope for SME companies to find appropriate opportunities and in the end cripples their chances for development as they are weakening the strength of their enterprise as if through ill-fated first-cousin marriages.

Regional hubs do not mean that power is devolved to regions equally from the governmental level, having regional structures compete for scare resources against one another, all offering the same product and skills and keeping any best practice they discover to themselves. Every hub of regional excellence across Europe should specialise and provide separate, extra-value partnering as huge innovative consortia with different types of excellence to offer, sharing best practice. This is the one way Europe can compete on a global level successfully, and this is how the Galileo global positioning system (GPS) project has moved forward so successfully.[4]

In fact, the interviews in Denmark brought out the trend for certain regions within one country to excel at a particular enterprise often for historical reasons. The strength of these regions should not be held back by local or national constraints but become enterprise hubs of excellence that can link across Europe superseding national boundaries and giving access to markets directly on a global level.

A well-established example could be the City of London, which is a particular region in London and England that is a world leader that rival markets have been trying to emulate. The City of London has the capability to continue in its path as a centre for excellence, a regional hub that continues to attract the best minds from the rest of world and as a global leader in financial services. The regional hubs of excellence could all be internally connected to each other like a digital city but on a grander scale than the trial cities found in other parts of the world such as the City of Sturt – a digital city within the City of Adelaide, South Australia.

This concept of regional hubs could be applied on a global scale, as the success of global innovation relies on attracting the best personnel, partners and expertise to make R&D successful.

Globalised innovation

As Peter Davies, CEO of Pera, has stated,[5] a passive rather than proactive form of globalised innovation is the most familiar model for companies. As large original equipment manufacturers (OEMs) have engaged in supplier development initiatives they have progressively whittled down the number of first-tier suppliers. Those surviving first-tier suppliers have had to take responsibility to deliver complete product sub-systems rather than just components, and that has put pressure on them to carry the main responsibility for innovation.

An example is car seating. Thirty years ago there were still first-tier suppliers able to make a business by separately providing seat structures, squabs, covering textiles, etc. (This will be explained further in Chapter 3.) Today, the seat is a complex electro-mechanical system, highly integrated, and in the next few years it will start to carry biometric sensors. The responsibility for new features in seating has largely passed from the vehicle OEMs to the first-tier suppliers, and from them partly down to second- and third-tier suppliers wherever they are located internationally. Thus innovation responsibility, and hence strategy, is becoming distributed globally.

There is however, a proactive form of globalised innovation that is in ascendancy. As Peter Davies emphasises in Chapter 3 of this book, companies that are at the forefront – and these have been initially high-tech firms – are taking seriously the facts that:

- the rate of product churn is increasing – for example, there are ever shorter life cycles for both manufactured and service products;
- the range of technology specialisations that can contribute new features to products is expanding rapidly and new emerging technologies are burgeoning;
- sustainable competitive advantage depends as much on business process innovation to support a manufactured or service product as on the product itself;
- no group of individuals or company has a monopoly on good ideas;
- almost all Western economies are short of skilled technological manpower.

The inexorable logic of these pressures is that, even if a company has the capability to handle its own innovation today, it is unlikely to have enough capability tomorrow. Some companies have explicitly encouraged their divisions to scout for relevant ideas and technology outside the company with the overall goal of increasing the amount of outsourced innovation from typically 10 per cent now to 50 per cent in the future.[6]

Thus the old model of expecting a corporate in-house team to be the majority provider of new innovation is moving to a new model of pro-actively building a global network of innovation partners. They may be suppliers, universities, other centres of excellence, and even competitors in certain circumstances. By aiming for the most appropriate and best partners, this network will almost certainly be international.

Each organisation brings with it a good knowledge of its local market, and thus the multiple and often hitherto missing linkages within the innovation network become a breeding ground for cross-fertilisation and a generation of new business opportunities, hence a ready mechanism for expansion beyond domestic markets.

Moreover there is the opportunity to break the bottlenecks caused by relative shortages of skilled manpower. The quality and numbers of good graduates in India, China, the Association of South East Asian Nations (ASEAN) and certain parts of Central and Eastern Europe (CEE) are major attractions when internationalising the innovation partners. Western economies, particularly the US, have in the past been very successful in attracting young graduates because of the career prospects offered. However, the rapid growth of business and state R&D in the rapidly industrialising countries means that there are now more domestic opportunities, and even if young graduates are still spending time abroad, there is an increasing return flow, as has particularly been seen with Chinese nationals into the special economic zones – a completed cycle of the global innovation.[7]

COMPANY CASE STUDIES

Piezotag Ltd

Piezotag Ltd is an innovation business partnering success story coordinated and run by Geoff Haswell, the managing director of the company. He believes that product innovation involves a new or modified product while process innovation concerns a new or modified way of making a product. Innovation sometimes consists of a new or modified method of business organisation. In many case examples, the introduction of the credit card has involved all these types of innovation. Geoff contends further that most innovation does not end in a new product, process or way of doing business as innovation is an almost seamless process, often leading to dead ends and places of no commercial interest. Sometimes its only benefit is the journey, the seeing of new things and the understanding of new processes. Something is gained but it is not easy to define or it may not seem useful at the time.

All three definitions of innovation applied to the creation of Piezotag, which was an idea that began with a meeting at Pera, which revolved around the use of piezoelectrically produced power to drive a remotely located sensor, and by using radio frequency transmissions to send the collected data to a display unit in easy reach of the operator. Geoff notes that the application was the result of the disciplined nature of the discussions, which in many ways had a spontaneous outcome that worked.

Minimising risk for innovators and entrepreneurs is best achieved by spreading the risk among many investors and many R&D performers. As Piezotag was born in the EU-funded CRAFT programme and because the programme allows for technical and commercial development of an idea or concept, it meant that by the end of the project, which lasted two years and three months, Piezotag had achieved a saleable technology. The role of an external innovation facilitator was instrumental, according to Geoff, as the facilitator had a laboratory with technologists who understood the technical – in this case, electrical – process, and it was Pera technologists who suggested the tyre application for the CRAFT project. Pera also filled in all the forms for the Framework Programme, for which an SME company does not possess the resources or time. Geoff came from a plastic moulding background, and required the assistance from Pera technologists regarding the electronic aspects.

The business partnering aspect was crucial to the success of Piezotag, with Pera putting Geoff in contact with a printed circuit board (PCB) motherboard company, another plastics company from Germany and the University of Catalonia, which opened the door to SEAT in Barcelona for potential commercialisation. Great trust was built from the point of idea conceptualisation to working with other consortium members through to the legal arrangements. Pera own 2 per cent of the Piezotag project, which gave them an added incentive to see the project through to fruition.

When it came to the point of commercialisation of the device, obtaining external funding was not easy. Geoff believes that in the final analysis it is the perceived commercial demand that dictates how easy it is to get funding. Commercial viability depends on whether the product will sell at a realistic price. Technical viability is determined by whether the product can be made at that price and patent protection is crucial, for there is no point in developing a product if you cannot protect it. Technical viability can be assessed by studying existing products from the appropriate market sector and matching their performance against the proposed business approach. If successful in obtaining external funding, it serves both as a technical and commercial validation of the product's success.

Through Pera, the University of Catalonia and Piezotag were put in contact with Renault and SEAT. Renault is already using a competitor's product, liked the functionality of Piezotag and will test it. Grant Thornton certified the Piezotag business plan and made it acceptable to investors who comprise a number of private investor individuals and two regional funds, the Advantage Early Growth Fund and the University of Warwick Concepts Fund. Both these funds are regional bodies centred in the Coventry/Solihull/Warwick area; their brief allows them to match pro-rata with private money up to their respective limits.

At the level of patenting, Piezotag shared the patenting costs across the consortium. This patenting pool situation ensured that all consortium members who assisted in the patenting process could only gain from the innovation they served to create. It reduced the financial burden as well as the risk involved in the patenting process. In the final analysis, Geoff found that being part of a wider European network had some distinct advantages. The EU-sponsored CRAFT programme is a good enabler, giving SME companies the wherewithal to undertake impressive R&D programmes outside their financial or technical ability. The only draw-back that Geoff found with the current EU programme is the dissemination of knowledge of what is available to assist SME companies. He would have liked to know more about what is available to assist companies in the US.

The Piezotag project was about as close as one can get to a complete success story, as the innovation facilitator, in this case Pera, was able to provide the right technical expertise, consortium contacts, successfully navigate the CRAFT FP6, use the consortium contacts to find future markets and outlets for the point of commercialisation, work together to find private and regional investors after assisting in the production of a viable business plan and, through the consortium, have the product patented before testing by external companies. The patent pooling served to spread the cost risk and drew the consortium members closer in trust towards their ultimate goal of launching a successful new product on the market that is in great demand.

Not all the companies interviewed had such a seamless success in business partnering.

The FIRICS project

The Fault Identification Retrofication Intelligent Computer Vision System (FIRICS) project was developed with seven industrial partners from four different European countries (see Table 2.1), each bringing complementary expertise with them. These complementary skills were enhanced by the R&D capabilities of Pera and Mindgroup, the innovation facilitators, who brought to the consortium the needed skills from their technology experts to achieve the critical developments in the project within the areas of computer vision and case based reasoning (CBR) software development.

Table 2.1 FIRICS partnership

A1	DKI Plast A/S	Denmark	End user
A2	Polyfa Trading A/S	Denmark	Injection moulding machine agent
A3	D-Codex OÜ – now Apprise OÜ	Estonia	Software development integration
A4	Industrias Tomas Morcillo s.l.	Spain	End user
A5	Image Scan Holdings Ltd	United Kingdom	Machine vision development and sale
A6	Plastindustrien I Danmark	Denmark	Trade association
A7	Delphis-IT Ltd	United Kingdom	Software training and sales
B1	Pera	United Kingdom	Computer vision and software integration
B2	Mindgroup	Denmark	CBR software and computer vision algorithms

The coordinator came up with the initial idea, as they were beginning to witness the effect of globalisation, and knew that to be able to continue with production in Denmark they would have to work smarter, as labour cost differentials made competition with the Asia Pacific and Eastern Europe unfeasible.

They therefore proposed to develop an autonomous, intelligent, hybrid data input, computer vision-based fault identification system to perform real-time product/part inspection and validation at key process stages to provide 24/7 product quality control (Fault Identifier; all terms here explained later). This was integrated with a low cost, intuitive, CBR-based diagnostic tool, which enabled low- to semi-skilled operators in diverse plastics processing environments to rapidly isolate and rectify the causes of faults identified by the Fault Identifier (On-line Expert). Underpinned by a case base of faults and solutions acquired from plastics processing experts affiliated to leading centres of excellence from across the EU representing commonly occurring, context-independent cases (Core Case Base), this was enhanced by an easy-to-use tool enabling low- to semi-skilled operators to augment the core base with context dependent, application specific cases (MyCase).

The project came to an end in October 2004, and has resulted in a prototype system that consists of two separate elements:

1   Fault Identification part – 'Fault Identifier': the Fault Identifier is an autonomous, intelligent, hybrid data input machine vision based fault identification system, enabling the system to quickly spot and identify a variety of plastic processing product faults.
2   Fault Rectification part – 'On-line Expert' consists of three tools:

   (a) MyCase toolkit, a case building toolkit that allows semi-skilled plastics processing operators and plastics machinery equipment maintenance personnel to register cases within the case base quickly and easily;
   (b) Core Case Base, a case base of commonly occurring, context-independent plastic processing faults;
   (c) On-line Expert tool, an intelligent CBR diagnostic tool that allows low- and semi-skilled plastics processing machine operators to query the case base quickly so that they can rapidly solve encountered problems.

The two separate elements can be brought to the market as an integrated product or as two separate elements as described above. It is estimated that the two separate elements will need an additional six to twelve months of development before they are at the point of commercialisation, whereby they can be marketed and sold effectively. Jan Zandhuis of 3DX-Ray, a subsidiary of Image Scan Holdings in the UK, believes that demand and necessity makes for supply and that successful product marketing is both market and customer driven. If a product is too new and the marketing has not been done methodically, people will not recognise the product's value. He notes, therefore, that a lot more assistance from innovation facilitators could be given at the level of the point of commercialisation and patents. Patents can be difficult because for protection they are necessary, but if done at a stage of rough ideas, others will understand the mechanics of the not fully developed product. For SME companies the biggest problem is once the prototype has been developed large companies, the usual targets of marketing for the product are only interested once the product has demonstrated its profitability.

Image Scan Holdings and Apprise OÜ will work on the prototype to bring the two separate developments to market. Innovation facilitators would be most useful for this stage of development.

Alari Aho, managing director of Apprise Estonia, was brought into the FIRICS project and will be working on the prototype with Image Scan because the CRAFT programme stipulated that a partner was needed from Eastern Europe. Apprise is a young SME company that has been operating for only four years. Alari was an entrepreneur for many years before that. Apprise is a software development services company that creates software programmes for such services as billing and document management and mobile solutions.

Alari believes that too much government support will make a SME company weak, which will kill the company when in the real world of competition. However, during the product development process, government support is essential, as it is too risky or even impossible for an SME to finance the innovation by itself. The CRAFT programme and consortium did not make Apprise reliant as it only received intellectual property rights (IPRs). In addition to the valuable IPRs, Apprise joined an expanded network of knowledge and gained valuable information concerning businesses in the UK and Denmark. No immediate profit came out of the project, but good technology resulted, training computers to recognise faults. A genetic algorithm was created for the computer so it can itself spot the nature of the fault, which it can then recognise in subsequent cases when working. The algorithm is copyrighted.

The problem for the prototype, according to Alari, is that it needs to be developed in a real factory situation where external factors such as noise, heat and other variables could impact on the genuine working of the product. Image Scan are keen to use the technology applied to their x-ray solutions for detecting explosives or other devices in luggage or baggage at airports and for food production such as detecting when cauliflowers have reached the right size and condition for market. Apprise, being a very young SME, does not possess the resources to commercialise and develop the product further. It is difficult to find pilot customers and unless the product has demonstrated its saleability, VC individuals or groups are not interested.

Apprise therefore sees a real need for innovation facilitators at the level of the point of commercialisation, serving to assist in finding pilot customers to further develop a good product such as has resulted from this CRAFT, Pera and Mindgroup assisted FIRICS project. They need to have a methodology and direction on how to prove that their product is a saleable one that can jump-start such innovative software programming for fault detection and other applications that involve the training of computers.

DKI, an end user of the consortium group, comes from another perspective. Although an SME and a budding mid-sized company, it is well established, having been started by one of the Neilsen brothers in 1946. DKI plastics manufacturing is a sister company, and they have three factories – two in Rosskilde and one in Jutland – and a new company in Slovakia. They also have a small representative office in Leicestershire. With so much diversity in interests and being family owned they are able to raise capital internally for new products. Despite this, the consortium idea is useful for them as they can avail themselves of new, well-developed products without prohibitive R&D costs.

Government funding and innovation facilitators working together – Enterprise Estonia

Egert Valmra, who was working as an expert at the Analysis and Development Unit at Enterprise Estonia believes giving funding alone is not sufficient and that the capability for innovation needs to be improved in Estonia especially in relation to application.

The EU provides structural funds of up to 75 per cent to match that of the Estonian government's 25 per cent funding. There is follow-up to assess the output and impact indicators as well as surprise audits. The amount of rules and bureaucracy can be overwhelming and Estonian ministers have not been able to find ways to ameliorate the rules, especially as Estonian law regulates how the structural funds are to be given out. Amendments to Estonian law were made in 2004. Egert noted, however, that external innovation facilitators have been invaluable, allowing Enterprise Estonia and companies being trained to make good use of the funding without the red tape and incompatibility with Estonian law to undermine innovation consortium projects from their inception.

Innovation facilitators such as Pera and outside consultants have helped to improve the application of innovative ideas and practices through raising the overall awareness of innovation and by delivering assistance to companies to learn innovation. Pera worked in conjunction with Estonian consultants to develop innovative ideas and computer strategies, and to provide the funding for projects. When the level of marketing the products developed with innovation consortium and facilitators is reached – at the stage of commercialisation – assistance stops. Enterprise Estonia then applies for some funding to aid these next stages through small grants from the Estonian government of up to 60,000 euros maximum. More assistance is needed at these stages with innovation facilitators providing guidance through training Estonian consultants at this level of expertise. A basic training programme with Pera already exists. The aim was to train sixteen consultants; fifty-four applications were received and there are now thirteen left on the programme. The potential consultants were chosen very carefully, with the majority having had extensive managerial and at least six or seven years' consultancy experience. Egert sees a need for such training at marketing, patenting and commercialisation levels, so the Estonian consultants can then train Estonian companies.

Despite reports recording the great successes of Estonian companies in adapting to new technology and the remarkable growth rate, Egert believes that the fifteen-year-old economy will face a slowdown if they do not pursue greater expertise in implementing innovative ideas. Labour costs are increasing between 7 and 10 per cent a year. Companies are increasingly unable to compete on prices, especially in the textile industries, which are in decline. Salaries are outstripping growth. Estonian companies are making the mistake of competing on price and profits are in decline. Egert believes the only answer is to compete in terms of innovative ideas and products.

One of the obstacles Enterprise Estonia faces is the complacency of many Estonian companies. Egert noted that companies have become complacent because they are relatively well-off and do not see the need to push themselves further. They have not experienced bankruptcy as yet and cannot see the pitfalls of profits stabilising and then plunging downwards if no extra efforts are made. Many companies cannot see beyond the huge profits that they made initially, from often simple market opportunities. When real market pressures set in and businesses begin to fail, lack of innovation and entrepreneurial knowledge will begin to show clearly. Enterprise Estonia is trying to circumvent the inevitable by having innovation facilitators such as Pera work in tandem with them to prevent this from becoming a widespread reality.

Egert sees that the fear of failure is militating against innovation and that companies need someone to believe in them to overcome such fears. Innovation facilitators such as Pera provide someone to lean on and show the way forward. Egert was attempting to develop more targeted programmes, finding the best people to train, what the emphasis should be, appraisal systems and how such projects should be managed in-house. He was responsible for setting up the information technology solutions.

Enterprise Estonia is far ahead of many other countries because its people already have an excellent grasp of the problems and are busy in search of solutions. While they recognise the importance of basic funding from the EU and even their own government, without an attitude shift by Estonian companies, which innovation facilitators can assist with, there will be no real progress.

Fear of failure and success in Denmark and Sweden

One of the common themes found in company and cultural attitudes when interviewing in Estonia, Denmark and Sweden was the how the fear of failing and being judged by colleagues and society as a failure has a profoundly negative affect when risk taking and being involved in risky innovation and entrepreneurial activity. All the companies expressed the view that they thought it was much easier to fail in one enterprise and then re-start another innovative project without being judged a failure in the United States. In fact in the US, failing and starting again is viewed as a natural part of the innovation process.[8]

Viewing failure as a normal part of the process allows entrepreneurs and company owners to remove a basic obstacle to success, for, as Geoff Haswell noted at the beginning of the company case studies, innovation is a hit and miss process whereby spontaneous ideas, after much discussion and trial and error, bring the desired result. An essential part of learning is trial and error so the fear of failure has more to do with culture and social stigma than reality. Egert Valmra noted that using an outside innovation facilitator can work to overcome feelings of personal failure as the external person can always be blamed for the failure, deflecting any social stigma away from the innovator who has failed by taking a high risk. It is preferable to create a culture of accepting failure and moving on, which exists in the US, but old, entrenched attitudes die hard so using external facilitators to promote new ideas and innovation reduces the risk of failure. They have accumulated experience, have been tried and tested and have successfully obtained Framework Programme awards for their clients countless times, which can only make the case for innovation facilitators for Europe a preferred option. Working with an innovation consortium is the other side of this coin because, as Geoff Haswell noted, risk in terms of cost and success is spread among a number of players rather than resting heavily on a single shoulder.

Fire Eater

Torbjorn Laursen, the managing director of Fire Eater, the Danish family owned company founded in 1974, is a genuine entrepreneur. He took over the family run company in the 1980s, in terms of part of the shares and the management, and turned the company around completely with his innovative ideas and entrepreneurial skills. Being a family owned company it was able to raise its own capital by using family assets as collateral. Any idea of a merger and acquisition was ruled out.

Torbjorn takes a hands-on approach, making it his business to know all fifty employees and what they do. He dresses informally, and the building is all on one floor so he can walk around easily. His office door is always open, and he has a lunch room for his employees that has high-quality, nourishing foods and a wide selection, and every week handmade cakes are made in the kitchen for everyone to enjoy. Those who work at Fire Eater are considered his extended family. He listens to employees, installers, plumbers and external customers. If an idea is good, he implements it. Torbjorn is also engaged in customer satisfaction analysis and responds to clients needs.

Initially, innovation and entrepreneurship were practised in almost a chaotic manner with brainstorming a random process, constant creativity encouraged and total flexibility. As the company has grown in size and profit, a quality control system has been instituted with regular product development meetings that involve, among other things, feedback from customers, discussion of products on the lists and marketing coordination strategies. There is an attempt to maintain an entrepreneurial spirit coupled with some discipline.

Fire Eater's central product is the epitome of innovation, coupled with social awareness about the environment. Fire Eater has pioneered the development of an efficient and environmentally neutral halon gas alternative to extinguishing fires. Torbjorn defined innovation as stumbling on ideas by coincidence inspired by searching for new standards and customer demands. Thinking of the alternative INERGEN, Torbjorn's secret mix of gases, which does not contribute to the destruction of the ozone layer or harm people, was an example of his preferred methodology. According to Torbjorn, while he was in the Middle East demonstrating the virtues of halon he became fully aware of its toxicological hazards. The halon gas decomposed in the flames into extremely toxic gases nearly killing him and the clients in the room.

Trial by fire is perhaps not the best manner in which to stimulate innovative ideas and is probably one of the most high risk, but it gave rise to a remarkable idea and the well-balanced product of the other chemical replacement gases of today, including HFCs and fluorinated ketones which are much worse than halon, and therefore are not an option for Fire Eater. It is simply too hazardous for people, and furthermore it is likely to destroy the assets it was supposed to protect.

INERGEN is unique because through its balanced mixture of the gases nitrogen, argon and carbon dioxide, it extinguishes fires by reducing the oxygen content in a room to below 15 per cent, the level at which most flammable materials can no longer burn. It is so physiologically balanced in terms of carbon dioxide content that the human organism is supplied with the necessary amount of oxygen despite the fire being extinguished through the reduction in oxygen content.

Despite his own capacity for innovation, Torbjorn needed to solve the problem of making less heavy cylinders to hold the INERGEN gas, but he had no expertise in plastics. He was put in contact with Pera Denmark and Lawrence Dingle, who has composite expertise. The problem was that non-magnetic storage containers made of a light aluminium were required to reduce weight for installation. Composite manufacturers that were contacted could not provide containers that both contained the INERGEN gases and were inexpensive. Lawrence suggested that an innovation consortium could pool resources to help solve this challenging technical problem. Torbjorn did not have the time or resources to apply for the CRAFT programme from the EU so Pera did this for him.

The result was the setting up of companies that already supplied or were customers of Fire Eater and Pera contacts from their networks. The innovation consortium consists of: D'appolonia, an Italian company that is an innovation facilitator like Pera that models software – in this case to calculate how to contain the gases; Technor, a Norwegian company that uses the Fire Eater product and has market contacts; Presto, a UK company that specialises in fixed installation and portable extinguishers; Coureasier, a French company that makes extrusion pipes to extrude the gases; and Eurocarbon, a Dutch company that deals with plastics.

This project is up and running, and Torbjorn believes it will have a number of positive outcomes. It will be important for the environment by making it more economical and preferable to use inert and non-chemical gases that do not harm the environment. A saving in transportation will result as less energy will be expended with lighter, plastic-based cylinders. For those who have to lift the cylinders, the weight will be less than 70 kilos per cylinder. A bonus for Fire Eater will be that it will control its own IP; it plans to set up a facility for manufacture in Europe and may license the plastic containers. Technor and Presto, which have access to markets, will assist with uses of the product.

Therefore, in this case the innovation consortium, with the assistance of the two innovation facilitators, Pera and D'appolonia, will be able to assist Fire Eater at all the different levels of product development, marketing at the point of commercialisation and with the IP dimension, including possible licensing.

Torbjorn noted, finally, that the innovation consortium was the most cost effective and valuable use of limited resources and compared it with using a university, which he had contemplated. To obtain such expertise at all the different levels of need, Fire Eater would have had to pay for all the wages involved, which would alone have caused him to require a 200 per cent reduction in overall wages. In other words it would have made this vital technology development beyond the means of a successful SME company such as Fire Eater.

Asah Medico – a company founded on advanced research

Producing state of the art laser equipment for partners in the US, Korea, Europe and Australia, the Danish company Asah Medico relies heavily on R&D. The company vision is to 'create advanced medical lasers for the benefit of the daily work in the medical field'. Olav Balle-Peterson, the Vice-President of R&D, explained that the privately owned company envisioned a new accessory for their laser systems with improved precision so lunchtime procedures could be carried out for busy people, who need face skin rejuvenation, hair reduction or vascular treatments.

In order to optimise the development of the new accessory for the laser system Asah Medico needed to complement their specialist expertise in medical laser systems with technical knowledge in areas of computer and vision technology. Asah Medico currently works with several US companies in very successful relationships, but despite this chose European partners for the project.

Asah Medico also has successful cooperation with universities (both university hospitals and technical universities). However, for this project Asah Medico was looking for a research partner with a more industrial focus. In addition, Asah Medico was seeking a more broad technical knowledge pool that could be accessed through the innovation consortium. To help build the innovation consortium Asah Medico was seeking an innovation facilitator in Denmark and found Pera.

After meeting with Pera Denmark in 2003, Asah Medico decided to move forward with an innovation consortium. Pera assisted with a CRAFT application, with the project beginning in February 2004. Through Pera's network of contacts they looked for potential partners with the specialist knowledge required that could work together to create a laser beam with intelligent precision and control. They needed research done to optimise the application so skin burning would be avoided while delivering effective treatment.

The combination of research requirements were difficult because sensors that respond to light and heat were required along with precise computer technology. In addition, all the processes involved were too potentially costly in time and money for Asah Medico to research by themselves. With Pera they established a bespoke innovation consortium to access European funding that could tackle this difficult problem. The consortium consists of: a supplier of camera platforms used for imaging infrared; a Swiss company that deals with mirrors and special coatings; a Dutch wholesaler for laser systems; a hospital in Sweden and a training centre in Poland together with a laser; and Fraunhofer and Pera for R&D and new technology development. For clinical trials Asah Medico will use a Danish hospital that they already know. This company shows that it is possible to advance using an innovation consortium despite the difficult nature of the advanced technology they are seeking to develop.

Asah Medico was founded in 1978, and since then it has supplied medical lasers and accessories. Despite being a successful SME company, it has been funded since 1997 by a group of public Danish investors who as venture capitalists and shareholders have provided an infusion of funds to the company to support the international expansion of the company and are still supporting it. This has ensured that Asah Medico has received the financial resources needed to strengthen its sales and marketing organisation.

Roxtec – even established leaders in their field need innovation consortia

Despite the excellent shape of the business, there was a substantial drop of sales of Roxtec in 2003. After some internal review and thought Roxtec realised the nature of the problem. By all accounts the Swedish company Roxtec has everything a mid-sized company could wish for. It was founded in 1990 by entrepreneur Michael Bloomqvist and produces a unique modular sealing system product that has modules that can be peeled back, making for flexible cabling multi-diameter technology. One box of modules makes installation easier and less heavy as well as less costly.

Joakim Hellkvist, Vice-President of Roxtec, noted that their products span between seventy-five and eighty markets worldwide and are three times the size of their nearest competitor. Roxtec has a 30 per cent growth rate per annum and just 180 employees. It also has affiliate companies in East Asia and Latin America that employ 400 people.

Joachim explained that the main sectors that they supply are the telecommunications, marine and industrial sectors and that their motto is 'We seal the world'. In true entrepreneurial spirit Roxtec views the world as one big cabling opportunity.

After analysing their internal problems they realised that they had lost their way through losing their focus on core values. The business had become top heavy with too many people involved in too much bureaucracy, which had caused the company to lose momentum. Joachim noted that they no longer focused on customer needs and had lost their responsiveness. They came to the conclusion that too many employees create too many people doing unnecessary jobs. With fewer people it is easier to maintain focus. Joachim believes that being a little understaffed creates priorities, but if there are too few employees, it is difficult to maintain quality. Responding quickly to the needs of customers is extremely important.

Roxtech attempts to standardise by having one face and one team to create brand development. Decision-making is very rapid and flexible. Everyone in the company is a decision-maker as it is delegated. All employees work on the same floor so everyone is aware of what their colleagues are doing, and if things go wrong, everyone must take a share in the blame. This last part of internal human resource reorganisation is to prevent a blame culture, which also incorporates the destructive force of envy, according to Joachim. This apparent European blight in attitude of blame and not wishing to accept responsibility for individual failure is an oft repeated theme in the countries of continental Europe, which is an impediment to innovation and entrepreneurship. In Scandinavia, people are protected from both success and failure through the extensive welfare state that tries to make all people the same. This tends to kill the entrepreneurial spirit and thirst for success, but helps to lessen the tendencies towards blame and envy.

The other element of getting the company back on its successful path, apart from the human resource reorganisation, lies with Pera to create an innovation consortium. This has begun with the Polish representative office, a rubber supplier that Roxtech already knows in Belgium and ABB, a large Swedish company that can help to provide suppliers. Joachim noted that Pera is serving to remove the burden of report writing, facilitate entrepreneurship and make available an extensive innovative network of 25,000 scientists, which offers substantial back-up.

Kryotrans – getting the foot in the door

Kryotrans, which constructs containers with on-board loggers for the distribution of refrigerated contents, had difficulties getting their foot on the first rung of the ladder to develop an innovative idea by John Pring, one of the founders of Kryotrans. John worked for the government on medicines control and then moved into medical logistics. He wanted to develop his idea of a container with an on-board logger to keep track of perishable goods such as medicines that need to be kept at a certain temperature for the duration of the journey to other parts of the world.

John found that the paper demands on SME companies through over-regulation makes for an onerous task. Kryotrans applied twice to the EU for a CRAFT award and were unsuccessful. They could not understand the requirements and turned to Pera for assistance, and after a second application of their own, Pera helped them to obtain a CRAFT award. John attended a seminar run by Pera on how to administer the European grant. As Kryotrans is very small and in the early stages with its new product, it did not have any extra resources to deal with the grant but only to give full attention to developing the unique logging system for refrigerated containers. John was able to delegate the bookkeeping and daily administration to Pera. All the consortium partners had a stake in developing the product.

The innovation consortium partners were as follows: Houghton Institute of St James Hospital Dublin – the R&D operatives; Olivo – a main partner from France who design and construct containers for refrigerated distribution; Vukov – a company in Slovakia with expertise in heat management; Clarepak – Irish plastic consultants from County Clare; Cybit – GPS specialists from Cambridge; and Biocair – a specialist logistics company for the pharma industry in Cambridge. The final team ended up as Olivo, who had been colleagues of John's for four years, Cybit and Biocair. John noted that Pera looked at every possible opportunity for team building.

With the innovation consortium's assistance, which was coordinated by Pera, the on-board logger product was developed successfully. Kryotrans, however, experienced real problems again at the point of commercialisation. They had the product but without adequate revenue streams. Nigel Brabbins, the CEO, obtained a grant from the DTI of £215,000. They also went to VCs, private finance houses and business angels to commercialise the product successfully. Nigel explained that it can take years to obtain vendor status, especially with pharmaceutical companies. SME companies can rarely afford market and sales research before proceeding. Companies that tendered to do this work for Kryotrans gave quotations for £200,000 upwards. This is compounded by the secretive nature of the pharma industries.

Nigel noted that to just qualify to be considered as a vendor, big companies expect quality systems, visiting those who supply them, grilling them on computing, requiring full documentation that can be reviewed at several levels and sending in their auditors. Kryotrans, therefore, believes that certain sectors are more difficult to commercialise new products than others. The pharma sector is a case in point, being conservative and secretive. This is despite prosecutions occurring for failure to control temperatures adequately in storage and distribution with spoilage of medical products resulting. John and Nigel think that innovation facilitators are very much needed at the level of commercialisation, where they can assist in opening channels of communication with pharma companies and expediting decision making. Kryotrans have no trouble in eliciting favourable responses from potential clients to the medicase product; it is turning that favourable attitude into concrete action that is eluding them, and what they need is assistance with this in addition to issues at the patent level.

AKI – nine dots – thinking outside the box

The first thing that Allen Green, joint Managing Director of AKI, spoke to me about was thinking out of the box, always looking for that extra bit of innovative value that exists in every potential project. Innovation in his opinion was not a new product but a new service that is rendered creatively, and some people are more creative than others. Innovations are good ideas that are creatively turned into products to serve customers.

With such fertile thinking Allen Green thought of three new innovative ideas and began seeking funding to develop them. Allen was attempting to diversify his plastic mouldings business because competitors were beginning to erode his market lead and create new ways to offer related products.

The realisation that he might be losing market share jolted him into new innovative thinking.

Allen was introduced to Pera through one of his major clients, Mira showers, as they had done an innovation project with Pera. Three projects emerged from these discussions:

1 A novel way of producing a shower plate using laser drilling that would make it more effective by revising the shower plate moulding: this idea was put forward to the CRAFT programme but was rejected.

2 A disposable insulin injection syringe made from plastic: Allen tried to replicate a steel needle in plastic made from strong polymer. There is a great need for a small, compact needle that is safe to use and easy to dispose of. The needle could either be burned or the end broken off. AKI – with a CRAFT award – conducted two years of research with Pera technologists but were unable to find a solution that would make the polymer strong enough to inject while being disposable. Limited resources have meant that Allen is still working on solving the technical problems in his spare time.

3 The Warmit project: this grew out of a project review initiated with Mira. It takes waste hot water from the shower to pre-heat cold water coming into the shower system. It was an initial idea from Mira, and Pera technologists added value to the concept by finding a solution that would prevent the product from clogging up the system.

The Warmit project was funded by the CRAFT programme and was an innovation consortium that included: AKI – a precision injection moulding company based in Hereford, UK; HRS Spiratube – a heat exchange company from Spain; Select Moulds – tool makers from High Wycombe, Buckinghamshire, UK – who organised the moulds; AST – a German company providing electronic measurement systems; Metallisation – a company based in Dudley, UK, that provides metal spraying equipment and services. All these partners put in time with Pera to develop the product technically. The route to market for this product is through Mira showers.

Warmit was developed successfully with Pera providing the box and AKI making it more attractive for sale. AKI (and Pera) has applied for a patent and is considering allowing Mira to obtain a licence to sell the product. A possible dispute could arise if AKI allows a rival shower company to sell the product as well.

Allen believes that the input from Pera as an innovation facilitator was invaluable for bringing the innovative ideas to fruition. He said that Pera projects help to change the culture of business towards invention and innovation through the methodologies and processes that they encourage businesses to follow. The networking opportunities are in themselves valuable and lead to more business through finding others to help develop innovative ideas.

Allen thinks that the CRAFT programme is lacking as it does not extend to assisting businesses at the point of commercialisation, which is the next most important step after creating and developing an idea. The lack of resources and reserves of SMEs and mid-sized companies means that they are stopped from progressing the innovation service/product as quickly as it needs to be. It would be useful to have an external facilitator to assist with the paperwork and marketing to promote the new product. With another product related to work by Pera and AKI, a patent dispute arose and this is an area where guidance from an external facilitator could be very useful.

Clinipart Ltd – a syringe in time

Clinipart Ltd are an innovation investor who look to develop new products in the medical sector. It is led by John Watkinson, an entrepreneur who invests his own money into projects that he feels have a social and economic impact. John's partner, Dr Joseph Peters, a surgeon based in Harlow, UK, is a key driving force for the new development of medical products. Following the death of a boy in a Midlands hospital, Clinipart decided that it was time to develop a new spinal injection system for the safe delivery of the correct drugs to the spinal cavity. To achieve this aim it was decided that the new spinal system should not be compatible with the existing Luer system currently used throughout the medical sector as the main connection system.

Following a meeting with Dr Ian McKay of Pera Innovation, Clinipart decided that the next step forward to achieve the new spinal system was to participate in an EC-funded project. This would gain them the necessary partners to form a complementary supply chain and research agencies to bring the concept through to fruition. Using Pera Innovation's expertise in gaining EC funding through the various thematic models and their network of SMEs, LEs and technology centres, Pera was able to put together a successful proposal for them.

To obtain the necessary funding for the continuation of this project, Clinipart, together with Pera, decided that the best thematic model to take this concept forward was that of CRAFT, a funding model that allows between 1 million and 1.5 million euros of research to be performed over a two-year period. Following the successful acceptance of the proposal by the EC, the project – now called ISOLOK – was started, jointly managed by Clinipart and the consortium partners: Dr Peter Kinast of Melab Germany, a renowned test laboratory for the current Luer system; Mr Robi Bernberg of Rocket Medical, based in the UK, a leading manufacturer of injection-moulded components for the medical sector; Patrick Mulholland of Vistamed Ireland, a manufacturer of drip sets and epidural systems; and Silcotech, a silicon injection manufacturer based in Switzerland. To complement these partners' skills, Clinipart also utilised Pera for its ability to successfully manage EC projects, for having the capability to deliver from concept through to finished product, and for having the technical infrastructure to deal with the legal issues raised by the EC during the course of the project. To assist in the delivery of the project with their knowledge of analytical techniques, Clinipart and Pera decided that the Fraunhofer Technology Enterprise Group (TEG) based in Germany would provide the best synergy.

The main concept of the project was to develop a new spinal system for the safe delivery of drugs to the spine; however, it was also decided that while the concept of drug delivery was being looked at, the project could also look at other problems currently contributing to the 10,000 litigations within the National Health Service (NHS) per annum such as disconnection and too rapid an injection.

During the course of the project, Pera helped Clinipart by developing a new spinal system that was totally incompatible with the current Luer system yet very much safer for the patient; the new system has an in-line connector that includes a unique ratchet mechanism to prevent accidental disconnection and a membrane technology for the prevention of too rapid an injection. All of these systems are currently on test with the NHS in the UK and have been disseminated at exhibitions such as MEDICA in Düsseldorf where the concepts received a great deal of interest from the global market.

Finsbury Orthopaedics – moving up to a mid-sized company

Finsbury Orthopaedics is in a high-growth area because as the post-war baby boomers are ageing they want to remain mobile, creating a greater demand for artificial joints. This growth in demand for medical devices is reflected in Finsbury's sales increase for the year to roughly £14 million. The company was started twenty-seven years ago as a university spin–out from Imperial College London by its founder Mike Tuke, with Bill Day and surgeon Michael Freeman. As the company has grown, they have developed a number of very successful devices, such as the Freeman Hip stem and the metal-on-metal resurfacing hip.

Dr Andy Taylor, a chartered mechanical engineer and currently the manager for the R&D department within Finsbury, explained two projects undertaken with an innovation consortium under the CRAFT award programme with Pera. He also noted that innovation is the lifeblood of the company.

The first project was the Hiped Hips II. This involved creating novel features on the surface of hip components to generate better integration between the bone tissue and the hip implants. The resulting integration means that the implant will bond with the bone tissues better and thus will increase the chances of early fixation of the implant, which, it is hoped, will increase the life of the implant within the femur. The novel features are generated by a process called layered manufacturing or rapid prototyping where a metal powder is thinly spread over the surface and then a laser melts the outline shape of the feature. This process is then repeated layer by layer until the structure of the surface feature has been manufactured. Once the novel features are created, these are coated with a material that promotes bone growth. The success of this project was helped by the expertise of Pera and their coordination of the project to the successful outcome.

As with all EC-funded projects a consortium of partners was put together by Pera using their extensive database of SMEs and where necessary their network of LEs. Pera were also able to bring another technical organisation from their innovation network to perform the *in vivo* testing that was outside the consortium members' expertise. The innovation consortium included: Tekimed SA – a French company that supplied and coated the implant with the bone promotion agent; Doxa AB – a Swedish company that also supplied and coated the implants with an alternative bone promotion agent; Haswell Moulding Technologies – a UK-based company that manufactured the mechanical test equipment and test sample for the validation of the manufacturing process; EOS GmbH – a German company that performed the laser melting process for the generation of the novel surface features; Goodfellow Cambridge Ltd – a UK company that supplied the powder materials for the laser melting process; Pera Innovations – a UK-based company that organised and coordinated the project with the consortium partners and the EC through Finsbury; Gothenburg University, Sweden, led by Professor Peter Thomsen, who performed the biological testing and acted as a second R&D performer to Pera.

The partnering for this project by Pera has been so successful that there is possible component work on the horizon with EOS, further research with Gothenburg University, where a summer student has a placement at Finsbury, and another project with Tekimed is being planned on a drug release system. Hiped Hips II was so well liked by the EC that they praised the project as one of the best they had seen.

The second project that in 2004 won another CRAFT award with Pera is 'Project RASPED'. Finsbury Orthopaedics intend to provide a surgical tool to reduce the risk of femoral fracture during prosthetic hip implantation, based on the RASPED instrumentation. This is a surgical rasp to be developed with Pera and a consortium of partners with expertise in the manufacture of the diverse components needed to both produce and drive the RASPED device. It is intended that the rasp body will be manufactured from cost-effective metallic material and that a series of pressed and ground teeth will be generated that will remove the material from the centre of the femur ready for the insertion of the hip stem. The driving mechanism behind the rasp body will be a series of piezo ceramics that will give the necessary force and movement to actively remove the trabecular bone from the femur. The more effective cutting and removal process of this rasp device and drive mechanism aims to negate the need for impact-based surgical methods and reduce the occurrence of femoral fractures.

Finsbury Orthopaedics is committed to a long-term view and invests approximately 20 per cent of turnover on R&D. Their marketing strategy, which began three and a half years ago, was to market products directly so that they could establish their own networks, and use this to introduce new technical development directly. Finsbury's objective is to become a leading implant supplier with a focus on long-term research to develop the next generation of implants.

Finsbury demonstrates clearly that innovation success is not just a hit or miss process but one of vision where R&D forms the background business allowing them to stay ahead of competitors through understanding what the market requires and taking a long-term developmental approach. However, even successful companies such as Finsbury Orthopaedics rely on well-targeted funding opportunities assisted by external innovation facilitators working in tandem with innovation consortiums – a recipe for business partnering success.

Notes

1   Parts of the report were (re)written in the following journals: Ruth Taplin 'Growing up big and strong; harnessing the growth potential of SMEs in the EU', *Patent World,* December 2005/January 2006, Issue 178, pp. 18–21; Ruth Taplin 'Managing funding for innovative European SMEs', IEE *Engineering Management,* February/March 2006, pp. 18–21; Ruth Taplin 'The future looks bright with the right partners', *Managing Intellectual Property,* December 2005/ January 2006, pp. 50–2; 'High-tech gets lion's share of gazelle cash', *IEE Review,* News section, November 2005, p. 15; Ruth Taplin 'Innovation business partnering in Europe and the USA – are European SMEs losing the battle?', *KnowledgeLink,* November 2005.

2   This refers to a number of SME companies, including some clients of Pera. The author has owned and operated an SME company for seventeen years, which won Exporter of the Year for the UK in 2000 for Trading Partnerships and Pathfinder.

3   From Masatoshi Kuratomi 'Intellectual property and bridging loans: their emerging roles in venture finance and business rehabilitation in Japan', in Ruth Taplin (ed.) *Risk Management and Innovation in Japan, Britain and the United States,* London: Routledge, 2005 pp. 162–77.

4   See Ruth Taplin 'Changing satellite systems in Japan within a global context', in Ruth Taplin and Masako Wakui (eds) *Japanese Telecommunications: Market and Policy in Transition,* London: Routledge, 2006.

5   See SME innovation report referred to on the opening page of this chapter (p. 9) and Peter Davies's chapter in this book (Chapter 3).

6   For a discussion of the management imperative to outsource see Alpesh Patel and Hemendra Aran *Outsourcing Success – The Management Imperative,* London: Palgrave MacMillan, 2005. See also Ruth Taplin (ed.) *Outsourcing in Japan, Europe and the United States,* London: Routledge, forthcoming.

7   See Ruth Taplin 'China and your career: threat or opportunity?', *IEE Student and Graduate Magazine,* February 2006, pp. 10–11.

8   See Terry Young's chapter in this book (Chapter 5) and Terry Young 'Technology transfer from U.S. universities: the need to value IP at the point of commercialisation', in Ruth Taplin (ed.) *Valuing Intellectual Property in Japan, Britain and the United States,* London: Routledge, 2004, pp. 20–33.

3   Globalised innovation

Peter Davies

Globalisation is a current vogue word for an old practice, encompassing the processes of increasing international trade, of reducing tariff and other trade barriers, of the increasingly free flow of capital, and of the building of controlled markets and alliances. Major trading empires have been created and destroyed by competition and warfare since time immemorial. It is true that, with the rise of communications and more pervasive media, the brand of a company has become more recognisable and this has given an easy point of reference for a popularist view that globalisation is a recent phenomenon. Yet starting in the early part of the last century, truly global businesses and brands have been built many times in transport, domestic appliances, processed food, media, information technology and communications, and latterly in leisure and sport.

There are well-proven strategies for companies to move beyond domestic markets into the international arena. The strategies have generally centred on:

- brand building;
- distribution and after-sales support;
- creating international supply chains for reasons of local supply and dealing with trade barriers;
- using international supply chains to feed global markets for reasons of cost competitiveness.

The strategies and the results are well studied and well understood. However, a new set of concerns now surrounds the term globalisation in the developed economies, and leads to urgent calls for governmental policy responses because of the pace and scale of three effects.[1] First, the digital revolution has created information and communication tools that in many cases remove the need for senior managers to be co-located with the processes they control. Second, the political triumph of the market economy, though not universally acknowledged, is now opening up trade and markets at a hitherto unachievable rate. Third, the large population centres of India and China, which dominated global wealth for centuries until the combination of the industrial revolution and military domination by the West overtook them, are re-emerging with huge domestic markets and labour forces at their disposal.

Thus much political debate currently centres on the opportunities and threats posed by the changes in employment and trade engendered by the sense of globalisation described above. This chapter, however, will go further and explore yet another trend, which is buried within the overall changes but is emerging as having particular significance for Western companies – that of *globalised innovation*, which can be a driver and enabler for a whole new phase of business expansion.

Globalised innovation

A passive, rather than proactive, form of globalised innovation is familiar. Large companies, particularly in the manufacturing sectors, have extensive networks of suppliers. The drive for efficiencies and cost controls has led to supplier rationalisation initiatives that have progressively whittled down the number of first-tier suppliers. Those surviving first-tier suppliers have had to take responsibility for delivering complete product sub-systems rather than just components, and that has put the pressure on them to carry the main responsibility for innovation.

There are many examples in the mature sectors such as the aerospace and automotive sectors. A case in point is car seating. Thirty years ago the main knowledge about the ergonomics of the driving position or passenger comfort and safety was concentrated in the corporate R&D centres and design centres of the large vehicle-assembling companies, that is, of the globally branded companies. There were many first-tier suppliers able to make a business by separately providing seat structures, squabs, covering textiles, etc., to the precise specification issued by the vehicle assemblers. The R&D of these first-tier suppliers, such as it existed, was mainly concerned with materials testing and validation together with manufacturing process innovations to keep their prices competitive. However, today the seat is a complex electro-mechanical system, highly integrated, and in the next few years it will start to carry biometric sensors. The responsibility for new features in seating has largely passed from the vehicle assemblers to the first-tier suppliers, and from them partly down to second- and third-tier suppliers wherever they are located internationally. Thus innovation responsibility, and hence innovation strategy, is becoming distributed away from the large corporate centres into the geographically spread supply chain.

Another example is computers. The first companies to enter the desktop personal computer market were themselves the drivers of most aspects of the technology. They specified their own computer architectures, they were actively developing their own mass-storage technologies in their large corporate R&D laboratories, and they either made their own processor chips or heavily influenced the specification of the chips fabricated by specialist suppliers. Today, the components of desktop and laptop computers have become commoditised. There are now separate global specialists in displays, processors, disc drives, other storage technologies, etc., and most of the innovation is being driven by them through research centres distributed globally.

The above two examples are of distributed innovation being driven, in a *reactive* way, by other business pressures. However there is a *proactive* form of globalised innovation that I would like to explore. Companies that are at the forefront in this arena are taking seriously the facts that:

- the rate of product churn is increasing, i.e. there are ever shorter life cycles for both manufactured and service products;
- the range of technology specialisations that can contribute new features to products is expanding rapidly, and new emerging technologies are burgeoning;
- sustainable competitive advantage depends as much on business process innovation to support a manufactured or service product as on the product itself;
- no group of individuals or company has a monopoly on good ideas;
- almost all Western economies are short of skilled technological manpower.

The inexorable logic of these pressures is that, even if a company has the capability to handle its own innovation today, it is unlikely to have enough capability tomorrow. Some companies, such as Proctor & Gamble, have explicitly encouraged their divisions to scout for relevant ideas and technology outside the company with the overall goal of increasing the amount of outsourced innovation from, typically, 10 per cent now to 50 per cent in the future.

This phenomenon in the large companies has been studied by, among others, Henry Chesbrough in his book *Open Innovation*.[2] His analysis contrasts the old paradigm of closed innovation:

> It is a view that says successful innovation requires control. Companies must generate their own ideas and then develop them, build them, market them, distribute them, finance them, and support them on their own. The paradigm counsels firms to be strongly self-reliant, because one cannot be sure of the quality, availability, and capability of others' ideas

with the paradigm of open innovation

> that assumes that firms can and should use external ideas as well as internal ideas, and internal and external paths to market, as the firms look to advance their technology. Open Innovation combines internal

and external ideas into architectures and systems whose requirements are defined by a business model. The business model utilizes both external and internal ideas to create value, while defining internal mechanisms to claim some portion of that value. Open Innovation assumes that internal ideas can also be taken to market through external channels, outside the current businesses of the firm, to generate additional value.

Thus the old model of expecting a corporate in-house team to be the majority provider of new innovation is moving to a new model of pro-actively building an external network of innovation partners. They may be suppliers, universities, other centres of excellence, and even competitors in certain circumstances. By aiming for the most appropriate and best partners, this network will almost certainly be international.

Each organisation in the network brings with it good knowledge of its local market and thus the multiple linkages within the innovation network can become a breeding ground for cross-fertilisation and generation of new business opportunities, hence a ready mechanism for expansion beyond domestic markets.

Moreover there is the opportunity to break the bottlenecks caused by relative shortages of skilled manpower. The quality and numbers of good graduates in India, China, ASEAN and certain parts of CEE are a major attraction when internationalising the innovation partners. Western economies, particularly the US, have in the past been very successful in attracting young graduates because of the career prospects offered. However, the rapid growth of business and state R&D in the rapidly industrialising countries means that there are now increasing domestic opportunities, and even if young graduates are still spending time abroad, there is an increasing return flow, as has particularly been seen with Chinese nationals into the special economic zones.

Experience of SMEs in Europe

If the logic of innovation by international partnering holds good for the large companies, it also holds good for SMEs who may need innovation more as a market differentiator in the absence of the international brands and market presence that large companies have.

To become global in one leap from an initial domestic market is a tall order. A few home markets, such as that in the US, are big enough for a company to grow to an internationally competitive size and then make a major investment in global expansion. But, for example, in Europe, trading within several countries is usually necessary to establish sufficient critical mass to act as a launch pad to global markets. The EU had, as one of its founding treaties, the establishment of a common market, but decades later there is still no equivalent to the home market in the US. The European market, though much more homogeneous than thirty years ago, is still fragmented by language, non-tariff barriers and consumer preferences. Companies struggling to innovate and bring new products to the market need first to overcome the fragmented market to build sufficiently large order books and investment to finance their business expansion.

This is one reason why the EC gives considerable support to transnational R&D through its Framework Programme. By grant-aiding industrial research only when it is collaboratively carried out or cooperatively commissioned by consortia of companies from several countries, it fulfils a role that no national government would undertake on its own. The new partnerships encouraged by the R&D subsidies not only make more innovation affordable in the short term but also build trans-European, business-to-business capability that will mature into more globally competitive strengths for the long-term future.

This is certainly beginning to happen at an increasing pace. Partly driven by the business pressures of product churn and emerging technology referred to above, and partly stimulated by the programmes administered by the EC, thousands of companies are now expanding their business outside their domestic market. My own company, Pera, has been at the heart of supporting this and, in fact, has facilitated the establishment of more pan-European industrial consortia for product and process innovation than any other organisation.

We estimate that over 1,600 companies have moved beyond their domestic markets, using an innovation consortium as the main vehicle, as a result of our efforts over the five years to 2005. And the pace is increasing, as twice this number of additional companies are expected to benefit in the subsequent five years. A few examples of recent product developments, drawn from the medical devices sector, will illustrate the transnational nature of the innovation consortia:

- an intelligent crisis-prevention management tool for asthmatics – companies from Denmark, Ireland, Israel, Spain and the UK;
- a new device for non-invasive blood glucose level monitoring – companies from Finland, Germany, Italy, Spain and the UK;
- the efficient manufacture of high-performance orthopaedic prostheses – companies from Austria, Belgium, France, Sweden and the UK;
- a fully integrated wound assessment and monitoring system – companies from Finland, France, Ireland and the UK;
- a safer system for spinal drug and anaesthetic delivery – companies from Belgium, Germany, Ireland and the UK.

Targeting resources

Most of the companies referred to above are experiencing innovation by international partnering for the first time. With their new partners they are creating market access abroad; they are generating new innovation networks that will outlast the first project, and they are building a critical mass that will allow them to plan for global opportunities. However, the budget allocated by the EC, and the effort expended by organisations such as Pera to facilitate and support such consortia, are both considerable and of premium value. It is therefore vitally important that both are well targeted on the companies most likely to succeed.

This has been well studied in a series of policy analyses from European bodies such as the European Association of Research and Technology Organisations (EARTO). John Hill, Managing Director of Pera Innovation, has collaborated with EARTO to develop the case that the majority of the European Framework Programme funding to SMEs should be focused on high growth potential SMEs.[3] High growth potential SMEs, often referred to as 'gazelles', were notably identified in the US by David Birch,[4] and are companies with at least 20 per cent sales growth every year.

The question is, where are the gazelles and what form of R&D and partnering support is optimal to assist them? Historically, European research policy has focused most R&D funding on SMEs that are high-tech research performers and, in the main, young and small. This is in part due to the common and widely held belief that there is a direct link between high-tech, young and small (between five and fifty employees) SMEs and high growth potential. Approximately 80 per cent of R&D funding for SMEs in the EU FP6 was focused on this typology of SME by channelling funding through the 'Thematic Priority Areas' where only high-tech SMEs capable of participating in research at the highest level of international excellence were able to participate. This policy has had the effect of concentrating SME research intensity into only the 2–3 per cent of high-tech firms capable of carrying out research internally at this level of excellence.

However, when the concept of the 'gazelle' was researched in the US there was little factual evidence found to support the myth that high-tech, young and small research-performing SMEs, in fast growing sectors, had the potential for high growth. Indeed, quite the reverse was concluded:

- Gazelles have little to do with high-tech, and the US figures suggest that only 2 per cent of high-growth SMEs are high-tech.
- Gazelles are somewhat older than small companies in general.
- Few gazelles were found in fast-growing sectors. Only 5 per cent of gazelles were present in the three fastest-growing US sectors.[5]
- Instead, the top five sectors in which high-growth SMEs were found (representing nearly 40 per cent of high-growth SMEs) were quite mundane and slow-growing industries such as chemicals, electronic and electrical equipment and instruments, paper products and plastics.

In 2004, Birch was called upon by *Fortune Magazine* as one of their 'Top Ten Minds' to revisit the subject of gazelles, and he provided the following analogy:

> Most people think companies are like cows – growing a lot when young and then very little thereafter. It's part of a natural tendency to ascribe a biological model to practically everything. And when we think of fast-growing companies, the tendency is compounded. We already know that rapid growers are small – and we are quick to equate small with young. It turns out we're mistaken. Companies, unlike cows, are regularly 'born again'. They take on new management, stumble onto a new technology, or benefit from a change in the marketplace. Whatever the cause, statistics show older companies are more likely to grow rapidly than even the youngest ones. Obviously, the high-tech boom produced a handful of very charismatic businesses. Truth is that most small business growth and most entrepreneurship does not come from the high-tech sector. It didn't during the boom and certainly does not today.[6]

Recent European research by the Organisation for Economic Co-operation and Development (OECD)[7] and others also confirms many of the US findings and observes that the negative impacts of globalisation on Europe's traditional sectors is creating the impetus for change in response to competitive threat, which in turn is driving growth among those able to respond with innovation. Some recent findings are:

- It is noticeable from the 2004 Europe's 500 that the profile of fast-growing companies has changed. There are fewer smaller companies, and older industry dominates the ranking. Fast and sustained growth is easier for companies operating in older, more established business sectors.[8]
- The highest number of champions of growth in the 2004 Europe's 500 is found within the manufacturing (industrial goods) and industrial services sector – 30 per cent of the five hundred companies – whereas the number in the information technology sector is down by half. Fast-growing companies can still be found in the technology sector, but they formed only 15 per cent of the Europe's 500 in 2004 (compared to 29 per cent in 2003).
- Maintaining existing businesses is a key challenge. An existing business has better chances of success and growth than a new business that has not yet overcome the typical 'teething problems' in its early years.[9]
- The percentage of high-growth firms in Spain was higher for companies aged 11–20 and 21–30 years than it was for those aged 0–5 years.[10]

This 'Great Gazelle Myth' that favours high-tech research-performing SMEs has become entrenched in European government support policies. The consequent relative marginalisation of medium-tech but highly innovative, entrepreneurial, older and larger SMEs spread across the broader population of sectors could also explain why American SMEs are between seven and eight times more research intensive than the European population. The FP6 largely ignored (on the basis of an 80:20 ratio of funding towards high-techs) the potential for medium-tech SMEs with gazelle features to be highly research and knowledge intensive, without conducting large amounts of research internally. The entrepreneurial nature and market focus of gazelles typically leads them to exploit outsource models for research performance, in which they use universities and research institutes to carry out the research to create the enabling technologies for their innovations. That is, they are more disposed to innovation by partnering at the outset, and they are more likely to be active participants in globalised innovation.

Further OECD research confirmed the link between innovative, medium-tech firms in traditional sectors and high growth potential SMEs, concluding that the balance of funding between high-techs and medium-techs should be redressed:

- First and most importantly, growth is closely related to a company's ability to innovate. Because high-growth firms are found among SMEs in all sectors, policies should address SMEs in a range of industries and regions; they should not focus solely on small high-technology firms – well-established and larger firms are important entrepreneurs.[11]
- There is a clear positive link between R&D effort and high growth, and product innovation plays a key role in the high-growth process. Although high-growth enterprises are more heavily committed than others to R&D, few have conventional R&D facilities. Those that lack a formal R&D department perform research in ways that depend more on networking and engage in partnerships with government, research or academic institutions.

In order to mobilise a greater range of SMEs from a broader range of sectors and with a greater potential for growth, it is essential that future government policies and, at the transnational level, future Framework Programmes redress the balance of funding for SME research. The 80:20 balance that currently gives preference to high-tech SMEs that perform research should be reversed to give preference to the medium-tech SMEs that are potentially as research and knowledge intensive and possess greater growth potential but exploit outsourced models of research execution.

The UK government has taken a very commendable position in stimulating the ability of SMEs and other companies, of all types and sizes, to benefit from international innovation partnerships. Through a series of support mechanisms under the banner of 'Global Watch',[12] the DTI provides UK companies with awareness of partnering opportunities, based on leading-edge technology from around the world. Assistance is available for companies to participate in technology-themed missions to targeted regions, to be helped with brokering services from international technology promoters (who are highly experienced individuals with both business and technology credentials), and to second a member of their own technology staff into an overseas partner company or to receive such a secondee.

A complementary programme called 'Global Partnerships', run by the organisation UK Trade & Investment, directs its assistance in the first instance at foreign companies wishing to find partnering opportunities in the UK, in many cases as a prelude to inward investment.

Making it happen

The position we have reached in this chapter is that we have discussed the nature of globalised innovation as a concept. We know that its increasing adoption by large companies, characterised by Henry Chesbrough's 'Open Innovation' paradigm, has been studied, and we have a better understanding of which smaller companies, particularly in Europe, are likely to be the gazelles and indeed to have a predisposition to innovation by partnering. But implementing an international innovation strategy is complex, so how in practice does it get started?

The establishment and management of an innovation consortium has some very different features to normal supply chain management, and there are pitfalls for a company inexperienced in this area. And as Henry Chesbrough commented: 'the paradigm (of closed innovation) counsels firms to be strongly self-reliant, because one cannot be sure of the quality, availability, and capability of others' ideas'. Any company embarking on the alternative, i.e. on a partnering approach, will have all these concerns at the forefront of its mind. Any lack of clarity on how to deal with these issues will create business uncertainty and it is this uncertainty that most often kills innovation at the outset.

In their existing domestic markets, most companies can develop value chains successfully, that is, the whole chain of relationships with their customers through to the selection and management of suppliers. Indeed, this is the core business process for all companies. However, innovation has subtleties. When new product development (NPD) is largely based on upgrading existing product lines, it is a clear and familiar process that can be specified precisely in all its stages and delivery goals. Business schools have been teaching NPD techniques and best practice for decades. However, the development of the next generation of products or services of sufficient ambition to be able to transform the fortunes of a company in the face of international competition involves much more uncertainty.

And if it is being done in conjunction with new partners for reasons of acquiring the necessary capabilities and resources, issues arise for a company such as:

- Now that the ambition of the innovation does not have to be limited by just the company's own budget, how aspirational should the programme be? This takes the issue out of the traditional process-driven realm of NPD and into true business strategy at the highest level.
- How can the market prospects, technology risks, timescales, competitor action, etc., be assessed for something that is a significant step forward on what currently exists?
- How should the vision be communicated both internally within the company and in a compelling form with prospective innovation partners who are going to be expected to share the risks?
- How can the appropriate innovation partners be identified and approached in the first place, especially if there are serious geographic and language barriers? Some of the innovation partners may be existing suppliers but many may not be – this will almost certainly be the case if the innovation step is significant.
- Do the new products or services need new routes to markets and, if so, are the appropriate companies also identified as part of the new innovation consortium?
- How should the exploitation rights of the new knowledge be shared and in what markets? Who will own the IP or take responsibility for protecting it? This will influence the choice of partners, whether the object is vertical integration – only involving companies in a coherent value chain – or horizontal integration – bringing expertise from different product sectors and opening up whole new market possibilities.
- How should the joint work programmes be structured? What should be the individual work packages and their inter-dependencies? What are the risk mitigation strategies to counter the increasing reliance on outside partners?
- How is the consortium to be managed?
- Will the company's normal financing and investment mechanisms support the new way of working? Is some degree of public subsidy now necessary to overcome an initial market failure?

The above points, and many others, are typical of the uncertainties facing a company that is willing to embrace innovation partnering but that has no track record in doing so. Pera has coached thousands of companies through the establishment of hundreds of consortium programmes and believes that an experienced facilitator at the outset can make the difference between seeing the challenges as too severe – causing a retreat from the process and no significant innovative step – and seeing the challenges as tractable – leading to a business-transforming innovation. The important role of a facilitating organisation has been independently researched by Professor Ruth Taplin in her 2005 report on innovation partnering for SMEs.[13]

The most important first step for the company, with or without the assistance of a facilitator, is to be clear and bold in its vision for what business transformation it is trying to achieve, and that means a thorough ideas generation process. Core to this process is the empirically determined observation that the vast majority of successful companies exhibit 90 per cent of their business value either within their customer relationships or within their internal business process (this being the technological or knowledge-based competences they possess within their business processes, manufacturing processes or product technology). Customers only buy benefits that are derived from advantages in product use, which in turn are delivered through specific product features. These product features are created through competences, and unless that feature or competence is unique and protectable, the differentiation and sustainability of competitiveness in the market place can be undermined.

Most mature, low- to medium-technology companies and service providers have few or no unique or differentiating competences, but they often have customer relationships that can be leveraged through innovation. Conversely, many high-technology, knowledge-intensive companies have significant levels of their business capital contained within their technological competence, but relatively little to no value in their customer relationships.

Both populations can be helped through the same innovation process, which in its simplest form, identifies new and differentiating competences for low- to medium-tech firms to exploit to provide their customers with higher-value products and processes, or generates ideas for new customers to buy differentiated products based on a high-tech company's platform competence.

The process that Pera has developed is based on a market-pull approach to innovation that focuses on seven fundamental buying drivers for any product or service. These are then related to product features and competences based on a rigorously developed and validated process that, step by step, relates:

- competences to benefits bought, through features enabled and advantages generated;
- benefits bought to specific product and service concepts;
- product concepts to consumer or business segments and markets;
- markets to market structures and chains of decision-makers within the structures;

- decision-makers to buying drivers and benefits bought, and hence marketing strategy;
- decision-makers to specific customer groups and individual customers, and hence sales strategy.

Facilitation of globalised innovation

The stimulation of globalised innovation (that is, proactive innovation by international consortium partnering) should be a goal for any public body concerned with the long-term competitiveness of its domestic businesses. Providing access to experienced facilitation and coaching is an obvious but essential first step. However, to capitalise on its benefits any support system put in place needs a number of features in order to overcome the current preoccupation with closed models of innovation, with a myopic focus on mainly trying to find local partners, and with the model that innovation is driven by technology-push. Despite their endless theorising about the nature of innovation, too many public bodies are still operating outdated practices that mainly give funding to the research and technology base, principally in universities, and then support 'technology transfer' activities in those same organisations. The result is supposed to be more industrial innovation but in fact it leads to a constant re-discovery that companies are not adequately taking up the proffered knowledge and are not sufficiently innovating in their products and services. This typically leads to redesign of the support schemes and yet more funding going into the supply side and technology transfer offices. This cycle of under-achievement of technology-push has been all too common through the last thirty years.

All the analysis and reported experience above makes it clear that companies can and do innovate when they are clear about their strategy, have a business-transforming idea, and are facilitated to negotiate a way through the resource constraints, i.e. direct stimulation of the demand side rather than of the supply side. Some of the characteristics of any successful facilitation or coaching system are:

- It must focus on a business-centric process in which individual companies are facilitated to create concepts for new products and services, which in turn will create demand for knowledge from national and international sources of technological enablement. This has the effect of vastly increasing the research and knowledge intensity of the industry base as well as the scientific knowledge flow within the economy as a whole. Because the strategy process must focus on an individual company and be largely carried with the senior staff of that company, Pera characterises this as an 'invasive' approach.
- It must utilise business intermediaries (distinct from technology intermediaries) to deliver the 'soft' innovation skills as part of a business

process. These actors should possess a track record in innovation and entrepreneurialism. A recent survey noted that only 8 per cent of SMEs will accept innovation or business advice from public sector intermediaries and that 52 per cent prefer such strategic advice from experienced practitioners or consultants.[14]

- It must provide a high level of integration between the innovation facilitators and the other existing business advice and technology transfer programmes in the appropriate region. Specifically, when following the ideas generation process and when planning how to acquire the innovation capability, it is important to integrate actors from other services, such as sector-aware business advisors, technology experts from academia and intermediary technology institutes.
- It must have a meaningful proportion of private sector finance contribution towards the funding of the enabling innovation programmes. In a demand-led innovation system, successful innovation programmes tend to be financed using public–private sector partnerships, in which companies invest in the support with which they are being provided, and their recognition of the value of that support is sufficiently high to enable all but the very first phase of 'need recognition' to be part funded by the company. This high level of value recognition in these services also helps to overcome potential public sector financing obstacles with regard to state aid rules that could otherwise restrict public sector support for interventions that are close to market, downstream from research and not seen as pre-competitive.

In addition, there are two further key features for a facilitation or coaching system for globalised innovation. First, it needs to have the capability to carry out extensive techno-economic assessments of new business ideas and to be able to research market data, competitor data and technology trends. This needs a multidisciplinary group of experienced analysts, which is generally beyond the resources of any other than the largest companies, once again highlighting the advantages of using specialist facilitation organisations.

Second, the organisation providing a facilitation or coaching system must be demonstrably well connected with international business. It must have its own presence in key markets or at least be collaborating with similar organisations abroad. In assisting a company to find the right international partners for a consortium, a great deal of the initial sifting can be done from public sources of information, from the internet and from specialist pay-for-use databases. But finally there is no substitute to a well-validated recommendation from an organisation that understands both the need and the facilitation process, and that has the possibility of first-hand contact with a prospective consortium partner.

In summary

Globalised innovation will increase as companies respond to the over-whelming business pressures of product churn, the emergence of new technologies and the shortage of key manpower.

Achieving globalised innovation is complex, but small companies can benefit as well as large companies. However, they often need experienced facilitators or coaches to assist them through the process. This is happening on an increasing scale in Europe in particular.

Experienced facilitation resources and other forms of tailored support are of premium value and hence should be focused on the potential gazelles capable of the greatest sustained growth.

The majority of gazelles are not high-tech start-ups. They are companies with excellent ideas and good management, in general in mature sectors, with a relatively long trading record. Though capable of utilising high technology, they are generally not themselves research-intensive com-panies and will be open to cooperative partnering.

Notes

1 See for instance the article by Martin Wolf 'A bigger playing field needs new goal posts', *Financial Times*, 20 October 2005.
2 H. Chesbrough *Open Innovation: A New Imperative for Creating and Profiting from Technology*, Boston: Harvard Business School Press, 2003.
3 FP7 Stakeholder consultation at Noordwijk, 13 October 2004.
4 David L. Birch *Job Creation in America*, New York: Free Press, 1987.
5 Bureau of Labor Statistics *Monthly Labor Review*, Washington DC: October 1995.
6 Chesbrough *Open Innovation*, 2003.
7 OECD *High Growth SMEs and Employment*, Paris: OECD Publishing, 2002.
8 *Europe's 500*, 14 October 2004, For the latest 2005 information see www.europes500.com.
9 Commission of the European Communities *Entrepreneurship in Europe*, Summary Report, 19 October 2003, Brussels: Commission of the European Communities.
10 Paul Schreyer, *STI Working Papers 2003 – High-growth Firms and Employment*, Paris: OECD Publishing.
11 OECD *High Growth SMEs and Employment*.
12 www.globalwatchservice.com.
13 Ruth Taplin 'Can Europe make it? SME innovation partnering – the missing links', Melton Mowbray: Pera International, 2005.
14 *European Trend Chart on Innovation*, Brussels: Commission of the European Communities, 2004.

4 Innovative practices in Poland

An organizing framework and action plans

Alojzy Nowak and Bernard Arogyaswamy

Technology and the state

Technology is both a driver and a product of the process of globalization. Clearly, when firms such as Procter & Gamble, Phillips, and IBM establish operations in more countries, performing functions such as marketing, production, and even R&D in their foreign locations, they leverage their own capabilities while enhancing those of the host nation.[1] Governments have, over the years, been cognizant of the importance of technology development to sustained economic growth. Most states have acted, through intervention, policy changes, or direct funding, to raise the country's technological profile. Japan, for instance, helped its firms undertake licensing of foreign technology, and later encouraged collaborative corporate efforts, often offering tips on products most likely to succeed.[2] Later, the Japanese made a mark for themselves either through product innovation – Sony, Canon, and Honda for example – or, to a greater extent, through process improvements resulting in higher productivity and better quality. Toyota, Matsushita, Suntory, and Tokyo Steel exemplify the latter group.[3]

Countries seeking to emulate Japan's success adopted slightly different technology strategies but ones that did not vary much from their role model's. South Korea bought technology from abroad with its export earnings, the state playing a prominent role by, in effect, creating the conglomerates known as chaebols, offering incentives for R&D, establishing science and technology parks, and so on.[4] Malaysia and Thailand, on the other hand, sought to lift themselves up technologically by inviting foreign investment, initially by the Japanese, but later by European and American firms, in order to capitalize on likely spillovers. Meanwhile India, after being frozen in an autarkic, import-substitution position for nearly fifty years, started emerging from its torpor in the early 1990s with a market reform program that remains hesitant at best.[5] China's moves have, over the past quarter century, resulted in large injections of foreign direct investment (FDI) and technology into the country. Both China and, belatedly, India have implemented policies aimed at fostering high-tech businesses by serving foreign customers through software complexes in southern India and in Shanghai, as well as through customer service centers to deal inexpensively with complaints/information needs of clients in the United States, Europe, and so on.

Technology in CEE: Poland's record

The transition countries of CEE have, over the past decade or so, made great strides to overcome their Communist heritage. The three largest economies that dominate the region—Poland, Hungary, and the Czech Republic—have attempted to accelerate the creation of a free market system through privatization, reforming the financial system, attracting large inflows of foreign capital, working towards EU candidacy and membership, and so on. In respect of technology the record in these three countries is, at best, mixed. Diffusion, it is clear, has proceeded apace, with consumer goods comparable to those available elsewhere. While only a relatively small proportion of the population can afford to buy what they desire (unlike, say, much of the Triad region, broadly covering North America, Western Europe and Japan), the latest in visual and audio equipment, automobiles, cell phones, and domestic appliances are freely available. The bulk of the R&D work for these and similar products is done abroad, with varying levels of absorption taking place. Era GSM, one of the leading cell phone providers, for instance, imports the bulk of its equipment needs, making sure it has well-trained personnel to maintain and operate it. Firms such as Whirlpool, Carrier, and Phillips have been very active in CEE but the extent to which their technology is dispersed into local operations remains limited. The fact that science and technology were given great priority and received relatively lavish treatment in resources during Communist times has proved to be a mixed blessing in countries caught up in the cross currents of globalization.[6] The focus of research efforts in the erstwhile Communist bloc nations was on select areas in physics and chemistry. While the work was generally of a high caliber, much of it was pursued within the ambit of the academy of sciences and universities. Nearly all of the research was government sponsored, was theoretical, and rarely had any market linkage.[7] Technology during the Communist era had worked itself, so to speak, into a "high-level trap,"[8] that is, through a continuing and additive process of "technology-push" it had reached a relatively advanced level in areas that did not require market validation. Equipment in industrial production, for instance, while not necessarily efficient or easy to operate, worked well enough to provide the outputs needed in the form of electricity, chemicals, plastics, or building materials. In the case of consumer goods, in the absence of competition, it mattered little whether the design, materials, or quality met customers' expectations or needs. Though much has changed since the collapse of Communism

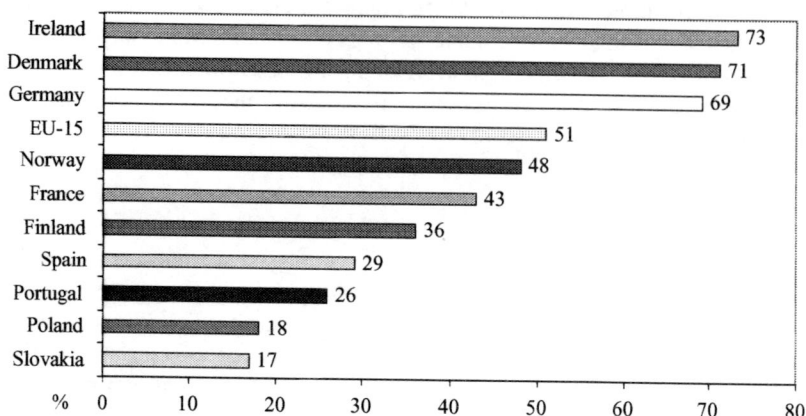

Figure 4.1 Share of innovation companies in production processes in selected countries

Source: "Innovation processes in Polish economy," Government report, April 15, 2005, Warsaw, Poland.

at the beginning of 1990s, the mind-set of research organizations, indeed, the culture of research, is proving difficult to change.[9] Compared with the EU average, the investment in and output from research activity in CEE appears rather low, with Poland faring worse than its neighbors in many respects.[10] In this regard, participation of innovation companies in production processes in Poland is a good example—see Figure 4.1.

In 2003, the number of patents per million in the population was 158.5 for the EU as a whole, 18.3 for Hungary, and just 2.7 for Poland. With regard to publications in technical journals (per million population), the corresponding numbers were 755, 370, and 221 respectively. The government share of R&D expenditure was 34 percent in the EU and 58 percent in Poland, the corresponding figures for the business share being 56 percent and 38 percent. The bulk of government R&D was conducted in educational institutions. Alarmingly for an EU candidate country, Poland's record on International Organization for Standardization (ISO) certifications till 1999 was dismal, just over 1,000 having been issued compared to its far smaller neighbors Hungary (3,280) and the Czech Republic (1,500). The number of internet hosts and personal computers per capita is among the lowest in CEE and the proportion of high-tech exports was the lowest at 2.7 percent in 2003, compared to EU-15 at 17.8 percent.

Technology in the EU

One of the expectations among policy makers in the EU candidate countries was that accession would result in a gradual or even rapid, convergence to

EU levels of economic performance and living standards. However, the potential for increasing returns to scale associated with high-tech businesses could lead to even greater technological (and economic) divergences between the present members and the newcomers. Considering that R&D expenditure constitutes less than 1 percent of GDP in CEE (0.7 percent in Poland), the EU average being more than 1.7 percent, the gap in absolute amounts expanded between the new and old members and is still monumental. To be sure, there are some ameliorating factors. The EU, for instance, has formulated long- and short-term technology/knowledge goals with a view to making the region a technology leader and to close the gaps with the United States (by increasing R&D funding to achieve the Lisbon Strategy goals).[11] Industries targeted for investment and, where necessary, state support, include biotechnology, ICT, and nanotechnology.[12] These areas, along with their support and complementary technologies, could constitute focal points for research in such countries as Poland with "catch-up" ambitions. In addition, the EU has instituted strategies and mechanisms to enable a coordinated approach to technology development across the 15, now 25, member nations. The Leonardo program, aimed at sharing information on building technical skills, and the European research areas (ERAs), charged with integrating research across the EU and with improving the linkage between research and market (the ERA funding alone amounted to 16 billion euros over the period of 2002–6), were among the many initiatives being pursued. Other projects include Eureka, in which transnational product development work was supplemented (to the tune of over 10 billion euros) and the Innovation Relay Centers (IRCs) in which transnational technology transfer and sharing of Eureka results were implemented.[13] Undoubtedly, after full accession to EU in 2004, countries such as Poland can, and must, make use of the resources and opportunities that are open to them now.

However, technology is not akin to a ripe fruit waiting to be picked, so to speak, from the EU tree. Much needs to be done by the new EU members. Moreover, if these governments adopt a "laissez-faire" stance or decide to intervene only in a market-failures scenario, they risk falling further behind advanced nations. Most governments in the latter are active in stimulating, if not managing, the continued growth of a knowledge-driven economy. The formation of a telecom cluster in a remote part of Sweden, the strongly directive innovation policy in Finland that resulted in one of the most wired societies in the world, and the Austrian push for high-tech business exemplify how relatively less populous states use government policy to achieve technological leadership. Even in the US, the bastion of free-market capitalism, innovation policy is an important contributory factor to the nation's technological preeminence.[14] True, a high-profile success story such as Silicon Valley owed little to state facilitation, but the establishment of science parks in various states, the extensive funding of academic research and incentives (such as the Bayh-Dole Act of 1980), and the designed spillover of defense R&D into civilian areas reflect a far from "hands-off approach" to innovation policy. However, in the transitional economies, such as in Poland, the stimulating role of government seems to be inevitable. Strategic push should be envisaged mainly in the free fields: education, science, and innovation policy.

A framework for technology development

Some of the actions that have proved instrumental in stimulating technology development and innovation in various countries are organized into the framework shown in Figure 4.2. We have divided technological capabilities into two categories—"innovation" and "absorption of innovation." The enabling forces, mainly exogenous to corporations, are broken down into those that are generated by the government, on the one hand, and by other support institutions essential to fostering innovative capabilities, on the other.

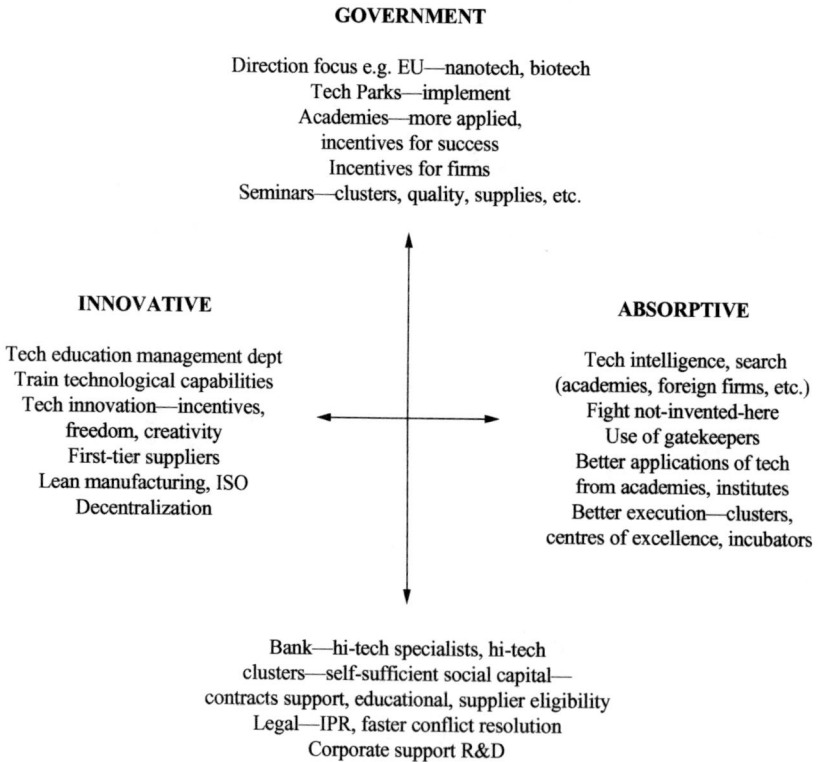

GOVERNMENT

Direction focus e.g. EU—nanotech, biotech
Tech Parks—implement
Academies—more applied,
incentives for success
Incentives for firms
Seminars—clusters, quality, supplies, etc.

INNOVATIVE

Tech education management dept
Train technological capabilities
Tech innovation—incentives,
freedom, creativity
First-tier suppliers
Lean manufacturing, ISO
Decentralization

ABSORPTIVE

Tech intelligence, search
(academies, foreign firms, etc.)
Fight not-invented-here
Use of gatekeepers
Better applications of tech
from academies, institutes
Better execution—clusters,
centres of excellence, incubators

Bank—hi-tech specialists, hi-tech
clusters—self-sufficient social capital—
contracts support, educational, supplier eligibility
Legal—IPR, faster conflict resolution
Corporate support R&D

Figure 4.2 Technological capabilities and their enabling forces
Source: The authors.

Some of the steps and activities that have proved effective in creating innovation development capabilities are:[15]

- the allocation of resources to technical education with a focus on disciplines central to the knowledge economy;
- establishing a context for learning in organizations, and the stimulation of creativity (including through the use of incentives);
- the development of managerial capabilities appropriate to technological advancement through greater decentralization;
- the development of lean manufacturing and high-quality facilities demonstrated by standardized certification, e.g. ISO;
- the ability to function as world-class manufacturing (and service) firms, or as first-tier suppliers to such firms;
- overall, a greater willingness and effort on the part of private enterprise to assume responsibility for innovation, reducing dependence on the state.

The ability to absorb technology from external sources, both foreign and domestic, is no less important than that of generating new knowledge. In fact, the two typically overlap and are complementary. Technology absorption can be improved by developing strong linkages with universities and research institutes to better utilize the research they conduct in the market by:

- developing more savvy technological intelligence and scanning to learn of developments in research institutions and competing firms (the use of boundary spanners facilitates this process);
- fighting the "not-invented-here" syndrome by stimulating a climate of learning from all sources;
- deploying gatekeepers to monitor and coordinate the transfer of technology to and from partner organizations;
- the formation of, and membership in, industry associations that set standards, enable knowledge transfer, and so on;
- participating in clusters, centers of excellence, and other such collaborative ventures.

Since research and innovation involve investments whose returns are uncertain and often not directly measurable, private firms may not be willing to pursue technology development. Such a "market-failure" phenomenon can be observed in less developed and transition countries but also in advanced nations as a matter of national policy or pride, and during difficult economic times. Japan's Ministry of International Trade and Industry (MITI now called METI) worked continuously and effectively over many years to facilitate the rise of Japanese industry by identifying potential growth industries, assisting in, even directing, technology transfer, promoting collaboration among firms and, in general, creating a competitive advantage for targeted industries. Other Asian nations, as we have seen earlier, learnt from Japan's success and adopted similar strategies to facilitate the success of local firms, and foster technological capabilities in the nation at large. As we also noted earlier, the EU's intent to enhance Europe's technological position vis-à-vis the rest of the world and the intent of individual countries within the EU to similarly forge ahead are clear and unequivocal. Among the steps undertaken by states to build a technological base are:

- providing funding to universities, research institutes, and corporations with resource allocation focused on specific areas e.g. biotechnology, ICT, etc. In 2006, at University of Warsaw in Poland, the Competition Committee for the Didactic Innovation Fund provided funding among others for such projects as "Information technologies and multimedia education in school practice," "Optical laboratory: opto-electronics studies," and "Laboratory of advanced measurement methods in physics";
- establishing technology and science parks with mechanisms to facilitate interaction and exchange;
- establishing agencies to transfer technology from research institutions to corporations;
- providing incentives for corporate R&D investments, for institutions to pursue innovations with market appeal, and transferring such innovations to firms;
- establishing or catalyzing the formation of clusters—for instance, a group of private firms cooperated together in generating new technologies in the aviation sector in southern Poland;
- conducting seminars on standards, quality management, ISO, contractual obligations, and so on;
- designating a single entity or agency, with appropriate authority, to formulate and implement all innovation policy.

While innovation policy and governmental efforts to sustain a national competitive advantage must be channeled through corporate efforts, the support of a raft of institutions is vital to the process being successful. According to the institutional economics perspective, changes in societal institutions are an essential prerequisite and concomitant to progress. Veblen[16] had theorized that technology is a process of social change, a dynamic factor, while institutions often resist change preferring the status quo. Any regimen of technological process must, therefore, encompass institutional reform or redirection.[17] Institutions critical to any program of continuing technological competitiveness include:

- financial institutions capable of evaluating, and willing to lend to, technology ventures bearing relatively high risk;

- legal and judicial institutions equipped to enforce laws swiftly and impartially;
- political institutions to enact laws and formulate regulations that will sustain innovation, e.g. IPRs, laws governing contracts, minimal entry and exit formalities, etc.;
- research institutions (including universities) with an increasing focus on corporate and market needs;
- educational institutions willing to alter content and shift direction in the delivery of technical and management education;
- social institutions that help build greater trust in society—industry and standards groups, training and skill enhancement organizations, and so on.

Polish achievements

Within this framework of technology and institutional development, we now examine where Poland stands in this context and offer suggestions on possible options that would facilitate far more rapid progress in innovative capabilities. Much has been done in Poland in the realm of technology policy during the 1990s, and much continues to be done today. In the area of technology policy, exemptions have been given to R&D organizations, and customs duty exemptions and provision for accelerated depreciation have also been made. Deductions of expenditure on capital equipment derived from R&D in Poland were instituted, but these were phased out in 2000.[18] It may suggest the lack of a long-term policy for science and technology, especially during the transition period.

Nevertheless, adopting a generally endogenous approach to innovation, numerous agencies and actions were initiated by different levels of government and by non-governmental entities. These include a "bottom-up" cluster in the plastic industry at Tarnów, technology transfer and information centers (to help with patenting marketing, business plans, etc.), centers for innovation to support new firm creation, business plans, marketing, etc.[19]

The funding for the academies of science and research institutes has been slashed, lowering the government expenditure on research and development (GERD) to Business R&D ratio (from 80:15 to 60:30).[20] Researchers in the academies now sometimes compete for funding based on project viability, which makes for greater purpose and accountability. The research institutes ("R&D units") are being consolidated and even partially privatized. Foundations and consulting organizations form the bulk of the non-profit organizations that have stepped in to support innovation with an applied focus. Technological parks, of which there are four, aim to create and help absorb technology while fostering regional development. Also attempting to accelerate regional development is a host of agencies (well over fifty). There are, in addition to these diverse groups, numerous other bodies, all of which work toward—or at least profess to work toward—various combinations of technology, business, regional, and other forms of development.[21]

It is clear that Poland has done much to build its technological capability, and it is continuing to utilize its tremendous assets to become an economic and technological power. Its history of small business formation (even during Communist days), high propensity for entrepreneurship, a rise in the population with tertiary education, the relative youthfulness of its population, friendly tax laws, and the multiplicity of agencies available to facilitate the creation and adoption of innovation are among the resources that contribute to Poland's considerable potential. While the strategies adopted and efforts expended all in the short span of a little over a decade are indeed commendable, some of the outcomes based on measures such as patents, ISO certification, articles published, internet hosts, and so on (as detailed earlier) are not encouraging.[22] Moreover, Poland is entering a new stage in its progression as a free market democracy and, indeed, as a country. While the country will compete no less intensely with other CEE nations for FDI, the challenges of EU membership, such as the easy access other EU-based firms will have to Poland's markets, must be addressed directly. Equally important, the opportunities provided by the increased access to new markets and technologies must not be missed or squandered.

However, the implications for new members of the EU are even more complicated. The problem involves imposing "high barriers to entry" to research participation by new member states. There are widespread fears even among the old and leading members of EU that the selection of a very small number of networks of excellence and integrated projects will spawn "self-reinforcing clubs," in which those countries and inner circles that are already well connected will become further entrenched and further supported. Along with high barriers to entry for new EU members go "high barriers to mobility."[23] Participation, for example in the widening range of mobility programs that are (rightly) on offer, often involves a ready availability of complementary resources to take full advantage, and too often these complementarities are lacking or deficient, e.g. at a personal level. The nature of "barriers to mobility" implies a path-dependent effect that persists in the absence of a disruptive force to break out of it.

Poland's technology options

Innovation policy—central coordination and focus

To begin with, the government needs to accept that it must take an active role in technology/innovation policy formulation, communication, and execution. Far from adopting a laissez-faire stance, the state needs to get involved in inducing businesses to undertake more R&D in their own interest, to facilitate (through incentives where possible) the pursuit and transfer of applied knowledge from the academies and institutes to firms, encourage developmental work in potential growth markets of "killer" products and so forth. In order to be able to plan and act decisively, it would be advisable for a single entity labeled, say, the Innovation Policy Committee (IPC) to be given complete responsibility for the coordination and direction of technology policy. The present system, in which the State Committee for Scientific Research (KBN) and the Ministry of Economy are charged with policy responsibility along with other ministries under whose purview the specific program may fall, results in no single entity being authorized to act toward or be responsible for policy success. There is no "owner" for any initiative. A body such as the Finnish TEKES,[24] reporting to a top official such as the Prime Minister, authorized and funded to make as well as implement policy, would be a step in the right direction. Establishing deadlines, timelines, and expected outcomes would further help put this activity on a "war" footing—which would not be an exaggeration of the urgency in the present context, as discussed earlier.

The predominantly endogenous approach followed earlier may also need to be eschewed in favor of a more pragmatic strategy. The desire and attempt to develop new technology by leveraging knowledge already available within the country is laudable, and may even lead to products/methods better suited to local needs. However, considering the rapidity of change in all areas of technology, it is critical that an efficient process of purposeful learning be instituted. The efficiency can be enhanced by reducing the duplication of resources, consolidating agencies, and so on, while the purposefulness comes from developing a focused approach to the technological effort.

A focus would constitute a single area or a set of areas in which the country would strive to achieve a distinctive competence. Examples would be communications, biotechnology, nanotechnology, etc. Rather than attempt to build capabilities across the board, an organized approach in a set of related technologies would help the R&D investment go further. This could be achieved by favoring academy proposals that are directed toward particular ends, by establishing technology parks and clusters oriented to the specialization, and so forth.

The discipline(s) attracting preferential resource allocation may be selected from the EU list,[25] particularly the ones for which the EU itself proposes preferential funding and intra-EU coordination. Linking up with ERAs, the IRC, the Eureka project, and so forth, would be a powerful means to gain traction in the road to technological excellence and competitiveness.

Technology linkages

Poland's technical capabilities, particularly in areas related to physical sciences, more so in basic research, are unquestionably strong. However, it has become clear in the post-Communist period that the linkage between scientific and engineering institutions and business firms is tenuous. In spite of all that has been done to tackle the disconnect, the ability of the researchers to address market or corporate needs and of the willingness of companies to tap into the pool of knowledge undoubtedly available at universities, academies, and research institutes, continues to stymie all attempts at creating a system of technological innovation. Several alternatives, not necessarily mutually exclusive, suggest themselves. One option is to institute a set of incentives—similar to those offered by the Bayh-Dole Act of 1980 in the United States—for scientists and academicians, whereby they are free to profit from their discoveries even if those discoveries result from government-funded work. Further down the "technology chain," the tax break on investments resulting from R&D performed in Poland, phased out in 2000, could be reviewed for reinstitution, coupling this with a tax deduction on investments resulting from industry collaboration with "upstream" institutions such as the academies and universities. While incentives might create a stimulus for increased cooperation, more will need to be done to overcome the inertia caused by a culture of "push" (researcher-driven) innovation, of absorption with pure research in the academies and universities, and of business skepticism of the fruits of collaboration with theoreticians. Liaison positions to establish links between business and the academy, and to uncover potential areas of common interest, would help transform the prevailing culture, as would participation in seminars, technology parks, and clusters. All of these need to be oriented around the areas of emphasis selected by the IPC and be part of units linked with EU research e.g. the ERA and IRC initiatives. While spontaneous cluster formation is laudable, in the context of countries historically conditioned to distrusting the unfamiliar, it is a slow process. Even advanced countries such as Sweden have acted to create clusters (in this instance, in telecommunications), and potential new entrants to the EU cannot afford to wait for initiatives to develop of their own volition and in their own country. Considering the vital importance of forging a vision and strategy for a national innovation system,[26] the actions envisaged in this chapter are of necessity of a centralized nature. This does not rule out action at the local/lower level by any means. However, technology policy should not be held hostage to, for instance, regional developmental needs. Establishing a cluster in a less developed region might be economically and politically advantageous but might be not be feasible in terms of attracting researchers, corporations, and infrastructural resources. Local authorities will, of course, remain at liberty to develop the institutions and resources they deem most appropriate, but these would not necessarily be awarded funding from the IPC budget. In fact, the proliferation of entities offering similar services (e.g. business planning) does not seem to have led to proportionate gains in new business formation/success. The IPC might consider disbanding agencies with overlapping action goals or establishing specialist agencies with purely advisory roles. For instance, the Center for Innovation and Entrepreneurship would function as a consultant to local-level bodies, a technology group would offer help on technical problems in business formation and operation, and so forth.

Complementary roles

Clearly, clusters, technology parks, small business advisory roles, and other similar strategies could help in stimulating technology development in a climate of better industry–academy collaboration. Small business start-ups, particularly in high-tech areas, however, also need access to sources of financing. While VC is available (though not to the extent needed), a real gap exists in bank financing for unconventional, moderately risky ventures. While most banks in Poland are modern in appearance and often in procedures (particularly the foreign-owned ones), they remain overly conservative and highly bureaucratic. Banks need to act to rectify the situation, and bodies such as the IPC could stimulate a review of lending policies both by interacting with institutional leaders and by their actions in helping develop more reliable innovation systems. Similarly, legal support for innovative activity needs to be bolstered. Manufacturers looking for first-tier suppliers, for instance, expect that contractual terms would be observed and are generally reassured by the existence of contract enforcement mechanisms. While the enactment of legislation to ensure adherence to contractual agreements would be ideal, in the shorter term, seminars and corporate training on functioning as extended arms of their customers, not merely as vendors, would be timely. Adherence to specifications and terms of contract are a small part of being first-tier suppliers. Joint product development and anticipating intermediate and ultimate customer needs are some of the responsibilities that go with this intimate form of outsourcing. More generally the importance of customer service in all sectors of industry as a complement to innovation has to be articulated and stressed. Since innovation is directed at users, the needs of the latter must be emphasized through the entire process of innovation (whether the customer is an individual or another organization), while skills at dealing with customers become almost as important as innovation capabilities for organizations and individuals as they get further down the technological chain.[27]

Technology absorption

Whether firms develop technology through transfers from the academies/ universities or by functioning as suppliers, subsidiaries, alliance partners of other firms, and so forth, the ability to absorb technology is a critical one. Certainly, some firms may be in a position to innovate without any assistance from external quarters, but the majority of Polish businesses are likely to be technology absorbers for the foreseeable future, due to the technological incongruity with the advanced nations. Improved, refocused education, retraining programs, and seminars could help create a greater ability to absorb technology. However, attitudes may also need changing. A preference for close supervision, higher than in neighboring Hungary, for instance, would have to be overcome both through skill upgrading and empowerment.[28] A tendency to distrust other firms and individuals, and the occurrence of the "not-invented-here" syndrome have to be dealt with. Management, traditionally centralized, high power distance (PD) and with low risk propensity has to adapt to the demands of innovation. Decentralization, employee involvement, the use of boundary spanners for filtering in outside ideas, gatekeepers for interacting with external entities including research groups, tolerance for mistakes, and incentives for innovation and adapting to customers' needs should become more the rule than the exception in managerial styles. In a sense business leadership in Poland has to run counter to the popular culture in creating innovative organizations, much as Akio Morita of Sony did in Japan and Ricardo Semler accomplished in Brazil. There are, of course, numerous types of innovations, not just the radical innovations that get much of the publicity—such as the laser, the internet, genetic modifications of crops and livestock, etc. Incremental innovations, while less striking, are by far more common and are both useful and lucrative. Process innovations, particularly in manufacturing, help improve quality and raise productivity as the Japanese have so often proved, to the chagrin of their competitors. Quality improvements fall into this category. Obtaining ISO certification is one indicator of process standards being maintained, and often serves as a potent marketing tool. Its absence, at any rate, serves as a deterrent to potential customers. Poland's performance on this count needs significant improvement. An inability or unwillingness to act on this could make it difficult for Polish goods to be exported even though Poland is a member of the EU. SMEs are likely to view the cost of certification as being prohibitive, which it well might be. If need be, a part of the IPC budget may be set aside for ISO subsidies and advisement, particularly since the demand for it is pressing, and the potential loss of market access nothing short of disastrous. Money is not the only bottleneck, however.

Corporations' ability to reinvent themselves to delegate more to better trained, motivated employees is at least equally critical to this initiative. The innovative capability of a firm also depends critically on its ability to recognize the value of new external information and to assimilate and apply it to commercial ends. Polish firms are often at the beginning of that process.

Conclusion

Just as corporations develop technology strategies and adopt competitive positions derived from their strategies, so do many states. Japan's initial emphasis on the absorption of product technology while aiming for leadership in process innovation paid rich dividends. The US government, while not overtly allocating resources to R&D, except quite substantially in defense, has funded research later transferred to private enterprise, The country's leadership in areas such as ICT and biotechnology has until now not been challenged. The EU, while bemoaning the widening gap with the US, has stated its intent to catch up through the funding, collaborative, and other strategies mentioned earlier.

In light of Poland's present technological and budgeting position, a strategy of "quick follower" combined with "opportunistic leadership" if, say, a focus on a branch of nanotechnology earns EU-wide recognition (e.g. in an ERA initiative), would appear to be most appropriate. Of course, the success of any strategy depends greatly on how well it is implemented. Resource allocation to select areas, a structure with a clear direction setter (e.g. the IPC) and built-in mechanisms for collaboration, coupled with a culture of market sensitivity, could be at the heart of a successful technology strategy for the country. Without a constant focus on emerging implementation problems, the national strategy would wither. Of particular significance is the political will to persist with future-oriented activities such as R&D in the presence of possible serious funding deficiencies for developmental projects with immediate tangible benefits. Creative approaches are required to fulfill the technological imperative. An example of this might be the contracts resulting from the offset (or the requirement to invest locally) included in the Polish government's contract with Lockheed for F-16 fighters.

Another example is the opening of Microsoft Center in Poland at the beginning of 2006 by software architect of Microsoft Corporation Bill Gates. The center will provide technical assistance to scientists and specialists in many areas. Information technology is a driving force behind economic growth in CEE, and Microsoft plans long-term investments in this sphere. The case of Microsoft Center is not quite typical. Instead of a brain drain policy, Microsoft prefers to be located in Poland, as a promising EU member, to build a solid, regional position. In addition, the offsets, which are local investments, may, with some imagination and perhaps considerable persuasion, be directed to allocation among different levels of technological effort. Borrowing from corporate technology strategy, Poland could build "base" capabilities in areas such as ISO certification expertise, "key" capabilities in, say, ICT, while "pacing" capabilities might result from work done in materials technology and, perhaps, microelectronics. Building technology parks and/or cluster around these offset investments would multiply benefits from the initial investment. (Base capabilities comprise technologies essential to be competitive in international markets; key capabilities comprise the core technologies or foundations for future growth; while pacing capabilities enable countries to achieve a competitive edge.)

Technology, though often viewed in terms of innovation in products, services, and improvements of processes, is in reality an abstraction representing change in human affairs. Veblen[29] and Hamilton[30] adopted the position that technology was a process of social change, and that institutional change touching all segments of society had to accompany the phenomenon of technology. While Veblen and Hamilton disagreed on the difficulty involved in institutional change, they are clear it is an evolutionary process. Institutional economics, which grew out of the efforts of these and other scholars, suggest that, without conscious efforts at transforming institutions, technology development efforts might not take root in a particular society.

In this chapter, we have proposed the launching of such an evolutionary institutional change process through the adoption of strategies and actions aimed at creating capabilities and changing existing mind-sets.

Notes

1   C. Bartlett and S. Ghoshal *Managing Across Borders*, Boston: Harvard Business School Press, 1998, pp. 3–5.
2   D. Okimoto *Between MITI and the Market*, Stanford CA: Stanford University Press, 1989, pp. 4–10; S. Callon *Divided Sun*, Stanford CA: Stanford University Press, 1995, p. 29.
3   F. McInerney and S. White *Beating Japan*, New York: Truman Talley, 1993, pp. 3–9.
4   C. Soon *The Dynamics of Korean Economic Development*, Washington DC: Institute for International Economics, 1994, pp. 12–20.
5   *The Economist* "A survey of India," February 22, 1997.
6   S. Radosevic "Introduction: building the basis for future growth-innovation policy as a solution," *Journal of International Relations and Development*, December 2002.
7   C. Nauwelaers and A. Reid "Learning innovation policy in a market-based context: process, issues, and challenges for EU candidate countries," *Journal of International Relations and Development*, Vol. 5, December 2002.
8   D. Boorstin *The Discoverers*, New York: Random House, 1983, pp. 3–9.
9   Nauwelaers and Reid "Learning innovation policy in a market-based context: process, issues, and challenges for EU candidate countries," 2002.
10 T. Mickiewicz and S. Radosevic *Innovation Capabilities of the Six EU Candidate Countries: Comparative Data Based Analysis*, London: School of Slavonic and East European Studies, University College, 2001, pp. 28–9.
11 EC www.ec.europa.eu, 2006.
12 M. Wolff, F. Kaehler, and C. Richard "Innovativeness and competitiveness among EU goals of knowledge economy," *Research Technology Management*, November/December, 2001, pp. 3–6.
13 Ibid.
14 G. Graft, A. Heiman, and D. Zilberman "University research and offices of technology transfer," *California Management Review*, Fall 2002, pp. 3–5.
15 K. Muller "Innovation policy in the Czech Republic: from laissez faire to state activism," *Journal of International Relations and Development*, December, 2002, pp. 18.

16 T. Veblen *The Instinct of Workmanship and the State of Industrial Arts*, New York: W. W. Norton and Co., 1964, pp. 28–30 (reprint).

17 R. Hollingsworth *Doing Institutional Analysis: Implications for the Study of Innovation*, Madison WI: University of Wisconsin, mimeo, 1998, pp. 24–6.

18 J. Kozlowski *Innovation Policy in Six Candidate Countries: The Challenges*, Louvain, France: Traverse d'Esope, 2001, pp. 6–9.

19 Ibid.

20 D. A. Dyker "The dynamic impact on the Central-Eastern European economies of accession to the European Union: social capability and technology absorption," *Europe-Asia Studies*, Vol. 53, No. 7, 2001.

21 A. Jasinski *Innovation in Transition: The Case of Poland*, Warsaw: University of Warsaw, 2002, p. 35.

22 Nauwelaers and Reid "Learning innovation policy in a market-based context: process, issues, and challenges for EU candidate countries," 2002.

23 N. Von Tunzelmann "Changes in the European system of innovation and the EU enlargement process" in Wydawnictwo Naukowe Wydziału Zarządzania (Publishing House of the School of Management) *Transition Economies in the European Research and Innovation Area*, Warsaw: University of Warsaw, 2004, p. 17.

24 Mickiewicz and Radosevic *Innovation Capabilities of the Six EU Candidate Countries*, 2001, pp. 28–9.

25 Wolff, Kaehler, and Richard "Innovativeness and competitiveness among EU goals of knowledge economy," 2001.

26 EC www.ec.europa.eu, 2002.

27 A. Havas "Does innovation policy matter in a transition country? The case of Hungary," *Journal of International Relations and Development*, discussion paper, December, 2002/5.

28 A. Laszlo and J. Fustos "Information sharing by management: some cross-cultural results, *Human Systems Management*, Vol. 18, No. I, 1999, p. 9.

29 Veblen *The Instinct of Workmanship and the State of Industrial Arts*, 1964, pp. 28–30.

30 D. Hamilton *Evolutionary Economics: A Study of Change in Economic Thought*, Albuquerque NM: University of New Mexico Press, 1970, pp. 4–7.

5 Innovation as the essential ingredient in American economic growth and future survival

Terry Young

Introduction

Building blocks for SME growth are many, yet most SMEs are fragile entities with vibrant commercial growth frequently out of reach for a number of reasons, including lack of continuous product innovation, failure to network internationally, regulatory barriers, outmoded technology and more. Much of the SME growth and success in the US is attributable to innovation, made possible not by some defining or unique characteristic of an "American personality" or "psyche" but rather by positive social and legal norms, well-defined systematic approaches to business and a vibrant government/business/academic partnership. As William A. Wulf, President of the US National Academy of Engineering recently stated: "There is no simple formula for innovation. There is instead, a multi-component 'environment' that collectively encourages or discourages innovation."[1] Furthermore, a robust business environment encourages SMEs to collaborate as well as to work with larger corporations to create domestic and international market opportunities that would not exist but through this collaboration.

A national innovation system

America's experience in economic growth since the conclusion of the Second World War has been examined in literally hundreds of scholarly books, empirical studies by academicians, and in the popular literature, some at times offering "diametrically opposed" explanations for the country's unquestioned standing as the economic leader of the world by the end of the twentieth century. Regardless, innovation is consistently acknowledged as a critical component of the nation's long-run economic growth. Dr Jeffrey D. Sachs provides one of the best histories of modern international economic growth in the 2005 book, *The End of Poverty: Economic Possibilities for Our Time*. In the chapter "The Spread of Economic Prosperity," Dr Sachs identifies the factors that have led to the phenomenal gaps between rich and poor countries. He concludes by noting in his summary: "I believe that the single most important reason why prosperity spread, and why it continues to spread, is the transmission of technology and the ideas underlying them."[2] The role of innovation in America's growth and the power of innovation is evident in America's invention, diffusion and commercial application of such technologies as the semi-conductor, the mouse, the internet, the personal computer, and e-commerce, all technologies that led directly or indirectly to the "knowledge economy" and contributed to the nation's business growth in the 1990s.

A misconception emphasized in much of the popular literature is that Americans are somehow unique in their zeal for innovation and resulting SME business development and growth. Simply as an example, in *Inventing America: A History of the United States*, the authors claim "America . . . was founded as a nation of innovation. The spirit of entrepreneurship is wholly in tune with our cultural, political and social commitment to innovation."[3] Notwithstanding, I personally can attest that Americans are by no means unique in their quest for entrepreneurial and even innovation-based commercial success. In my travels, I have encountered some of the most entrepreneurial and innovative individuals on the planet on the streets and in the shops of the least developed countries of the world. Dr Sachs suggests "the process of massive investment in research and development, leading to sales of patent-protected products to a large market, stands at the core of economic growth."[4] In a landmark cross-country empirical study of 2002, "The Determinants of a National Innovation Capacity," authors Furman, Porter and Stern describe this phenomenon as "the ability of country—as both a political and economic entity—to produce and commercialize a flow of new-to-the world technologies over the long term."[5] The balance of this chapter will seek to describe important components of this capacity and a "national innovation system" that led to America's economic success in the latter half of the twentieth century: (1) social views and cultural norms; (2) a strong IP regime; (3) strong investments in R&D; (4) strong government investment in SME support programs; and (5) ease in and strength of government regulations supporting SMEs. The chapter concludes with observations regarding economic realities that demand change for the twenty-first century.[6]

Social views and cultural norms[7]

Competitive success

Americans place high value upon competitive success in all facets of life, from sporting events to "spelling bees" to business competitiveness. In this fiercely competitive environment, SMEs are viewed as the "stars" or heroes of the business world, created and managed by risk takers, held by society in high esteem as those with the "gumption" or resolve to take on the many challenges of keeping a small business alive, vibrant and ultimately leading it to its economic success. Individuals such as Bill Gates, Steve Jobs, Sam Walton, Michael Dell, William R. Hewitt, and David Packard are just a few of the most well-known individuals who began their pathway to success as entrepreneurs creating SMEs at high risk. This spirit of competition and individual drive in business translates into innovation and real corporate growth in the US. Recent studies have shown that the new, rapidly growing SMEs ("gazelles" or firms that have annual sales growth in excess of 20 percent for four years in a row, from a minimum base of $100,000) are responsible for 70 percent of the net new jobs in the US.[8]

Failure in business accepted

Regardless of the high cultural value placed upon SME success, failure in business in the US is easily forgiven and often forgotten with no negative stigma or cultural consequences. In fact, failing in a first attempt to create an SME is something of a "badge of honor" or a "badge of courage"; I can envision individuals proudly wearing a T-shirt or cap with a slogan emblazed, "I Survived My First Failed Company!" The US Small Business Administration (SBA) found that 53 percent of small business persons who failed in their first company turned around only to form a second company to "try again."[9] It is often said, "I would rather hire a manager that failed in his/her first attempt to start a company than someone who never tried." It is hoped that a first-time failure teaches the many pitfalls and mistakes to avoid the second time around, leading to successful SME formation and growth. The adage "try and try again" is certainly applicable to the innovation-driven SME environment in the US. From personal experience, this is not true in most countries of the globe.

Open assimilation of new groups into the American business culture

In the US, there is an open and strong willingness to assimilate new groups into the high-tech SME business dynamic, and the country has benefited tremendously from this cultural norm. In fact, since the conclusion of the Second World War, America has been the beneficiary of an influx of many of the most talented minds on the planet. In *The World is Flat: A Brief History of the Twenty-first Century*, Thomas L. Friedman illustrates how the US benefited from the talented minds graduating from the seven Indian Institutes of Technology (IITs) established in the 1950s, with the best and brightest graduates "funneled" from India to Palo Alto, California, over a period of twenty-five years, fueling the tremendous scientific, innovative, and economic growth of the Silicon Valley.[10] As examples, a few of the individuals from other countries who were assimilated into the US innovation ecosystem include Vinod Kholsa (co-founder, Sun Microsystems), Jerry Yang (co-founder and Director, Yahoo), Sergey Brin (co-founder and President, Google), and Andy Grove (co-founder and Chairman, Intel).[11]

A strong IP regime

From the founding of the United States in 1776, the protection of IP has been one of the nation's fundamental foundations for innovation. Protection of IP was first set forth in the US Constitution, which gave to the Congress the power "[t]o promote the Progress of Science and useful Arts, by securing for limited Times to Authors and Inventors the exclusive Right to their respective Writings and Discoveries."[12] The first US patent was granted in 1790 to Samuel Hopkins of Philadelphia for "making pot and pearl ashes"—a cleaning formula used in soap making. From this humble beginning, the nation has been depending upon a strong regime for protection of IPRs to promote the advancement of science and the commercialization of innovations. IP protection for an invention induces investment in the development and commercialization of the subject technology. Generally, the US IP system supports the owner of the patent, copyright or trademark, thereby reinforcing innovation and providing strong incentive to SMEs to innovate. The IP system also encourages R&D, as SMEs may proceed with confidence that investments in the development of their innovations to marketable products are protected through the nation's IP system (including a supportive judicial regime).[13]

Strong investments in R&D

Public sector research

A 2005 OECD study found that public sector research is a primary "determinant of innovativeness in a country,"[14] while another study found a significant positive correlation between public funding of R&D and the "proportion of firms that are innovators and also with the share of turnover accounted for new products."[15] Undoubtedly, the federal government's role in supporting the R&D enterprise in the US is critical to the nation's innovative capacity and related economic growth. The government supports the majority of the nation's basic research—such as in biotechnology, information technology, and nanotechnology—the ultimate source of new knowledge that drives the innovation process and that will advance society in the future. Federally funded research also contributes to the education of the next generation of engineers and scientists; 59 percent of the R&D

performed at American universities in 2003 was funded by the government under competitive solicitations as well as through direct funding to Industry/University Cooperative Research Centers (I/UCRCs).[16] The internet, the web browser, bar codes, fiber optics, routers, Doppler radar, speech recognition software, computer aided design (CAD), nanotechnology, magnetic resonance imaging (MRI), the mouse, and GPSs are just a few of the thousand of innovations that have emerged from federally funded R&D in the US.[17] In December 2005, the US Senate introduced the NII, which would assign to the National Science Foundation (NSF) a central role that would "boost US competitiveness" through R&D. Supporting the critical role that innovation plays in the US economy, the language of the proposed legislation asserts that "research is *the key* to a strong US economy."[18]

As stated, much of the federal government's investment in R&D is made at research universities. The value of this investment is realized in many ways that drive innovation, in addition to the education of the next generation of scientists and engineers. For example, under the auspices of the Bayh-Dole Act of 1980, the title of the universities to their own research results funded by the federal government, and subsequently transfer those results to industry, further driving innovation and economic development. In *AUTM: US Licensing Survey FY 2004*, the Association of University Technology Managers (AUTM) reported $41.245 billion in sponsored research expenditures at 192 reporting academic institutions in 2004; of this volume, $27.72 billion was sponsored by the federal government. In addition, in 2004, responding universities reported:

- 567 new commercial products introduced to the market under license agreements with corporate partners (3,114 new products have entered the market under university–industry license agreements in the US since 1998);
- 4,783 new license or option agreements executed with corporations, with 27,322 agreements active, 22.4 percent of which resulted in running royalties on sales of products in the market in 2004; and
- 462 new companies based upon academic discoveries during the year (AUTM reports 4,543 new companies formed to commercialize new academic discoveries since 1980).[19]

The contribution of federally funded academic research is also realized indirectly in innovation-driven economic development. The Milken Institute has reported that "twenty-nine of the top thirty fastest-growing, high-technology metropolitan areas in the US are home to, or very near, a research university ... the availability of research centers and institutions is undisputedly the most important [location] factor in incubating high-tech industries."[20]

Private sector research

Sixty-three percent of the nation's $284 billion investment in R&D in 2003 was made by the private sector. Industry's share of the nation's research enterprise has grown steadily since 1980. From the end of the Second World War to 1980, the federal government supported the largest share of the nation's R&D; since 1980, industry as led the way in research volume.[21] The rate of industry investment (as a percentage of gross sales) is among the highest in the world, surpassed only by the Scandinavian countries. This investment is directed toward applied research (the "development" in R&D) and is perhaps the strongest engine for innovation-led economic growth and development.

Strong government investment in SME support programs

Federal government programs

The US government provides development funds to SMEs to complement private equity markets, fostering both innovation and SME growth. Some of the government programs for direct support of SMEs are described below.

The Small Business Innovation Research (SBIR) program provides federal R&D funds directly to SMEs "to improve the competitive capability of small R&D businesses with particular emphasis on emerging and underserved small firms," funding the critical start-up and development stages and encouraging the commercialization of new technologies, products and services. The program requires all federal agencies with annual extramural R&D budgets of more than $100 million (a total of ten agencies) to set aside 2.5 percent of those funds to competitively sponsor innovative advanced research within SMEs (businesses with less than 500 employees). Funds are awarded in three phases—Phase I for exploratory research (up to $100,000), Phase II to perform development work and evaluate the commercial potential of the subject technology (up to $750,000), and Phase III to move the innovation from the laboratory to the market. All SBIR awards are made through a competitive solicitation process. Since its establishment in 1982, the SBIR program has become the nation's single largest source of early stage research and technology development funding; in 2004, more than $2 billion was granted to 6,348 SMEs with "no strings attached" by the ten SBIR-funding agencies.[22]

The Small Business Technology Transfer Program (STTR) funds partnerships between small businesses and non-profit research institutes, including universities. Five federal agencies are required to allocate a portion of their budgeted funds for STTR, with competitive awards made in three phases, similar to the SBIR program. The STTR program is an effort to "tap research institutions for the enormous reservoir of ideas that have not yet been deployed effectively for the nation's economic benefit," funding research at SMEs to test the feasibility and scientific merit of new technology and to develop the technology to a point where it can be commercialized.[23] As an example, STTR might fund an SME to partner with a university researcher to spin off a commercially promising idea. In 2004, STTR awards totaling $209 million were made to 842 SMEs.

The Advanced Technology Program (ATP) of the National Institute of Technology (Department of Commerce) offers direct project grants up to $2 million for up to three years to US small businesses to "bridge the gap between the research lab and the market place, stimulating prosperity through innovation. ATP's early stage investment accelerates the development of innovative technologies that promise significant commercial payoffs and widespread benefits for the nation." From its beginning in 1990 through 2004, National Institute for Science and Technology (NIST) had made 1,511 awards to SMEs totaling $2.3 billion. Just as with SBIR and STTR, awards are made through a competitive solicitation process; cost sharing is required of industry awardees.[24]

Other federal programs directly supporting SMEs and innovation include the following:

1   The Manufacturing Extension Partnership (MEP) was created in 1988 as "a nationwide network of resources (more than 400 MEP offices) transforming manufacturers to compete globally, supporting greater supply chain integration, and providing access to technology for improved productivity." In 2003, MEP centers assisted 18,422 SMEs through technical assistance in advanced manufacturing techniques, training, and assessment of production processes.[25]
2   The Experimental Program to Stimulate Competitive Research (EPSCoR) of the NSF is an innovation promotion program established in 1978 under which NSF must set aside funding targeted directly to academic research institutes in nineteen of the fifty states that have historically received less funding than others.[26]
3   The Small Business Development Center (SBDC) program was created in 1980, and provides management, technical and research assistance through more than 1,000 offices across the country to aid start-up expansion and successful operation of SMEs trying to "foster economic growth through job creation and generation of new tax revenues."[27]
4   The incubator development programs of the US Economic Development Administration (EDA) have provided funding for the "bricks and mortar" of more than 600 high technology incubators in the US.[28]
5   There are more than 1,000 I/UCRCs, public–private partnerships with roughly one-third of I/UCRC funding provided by industry, one-third from federal sources (NSF, Department of Defense, EDA and other agencies), and one-third from state and/or internal university funds.[29]

State and local government SME support programs

By the 1990s, even when compared to the massive investments made by the federal government in direct support and grants to high-technology SMEs for innovation development and economic growth, the fifty states had emerged as even greater supporters of regional economic growth through the direct support and funding of innovation. Numerous catalogues and directories of "state innovation and economic development programs" have been published, seeking to list the thousands of state support programs. State programs are focused upon such initiatives as funding of R&D in "state of the art" technologies (information technology, biotechnology, nanotechnology, etc.), encouraging entrepreneurship and business start-ups, creating new forms of development financing for SMEs, stimulating academic technology transfer to SMEs, fostering expansion of existing SMEs, and promoting business investments in distressed regions and communities. In a recent comprehensive empirical study of US economic development, Dr Mikel Landabaso concluded that the *most inspiring lesson* from recent US experience is the enabling role that state and local governments have played in economic development through the promotion of innovation.[30] Another study has described the phenomenon as a "virtual incentives race" with states and local governments investing hundreds of millions of dollars annually in competition with one another to recruit or create the most innovative and leading-edge high-technology companies to their region through a host of funding mechanisms, tax credits, trade zones, and other creative initiatives, in themselves comprising a field of innovation in business and economic development methodologies.[31]

Ease in and strength of government regulations supporting SMEs

National governments have a significant impact upon the business environment and innovation, both positively and negatively, through policies that influence innovative inputs and development, and create the context for firm competition, demand conditions, tariffs, and many other aspects of a nation's business ecosystem. The World Bank's recent report, *Doing Business in 2006*, concludes that countries with simple and straightforward business regulations create more jobs than those with complex regulatory systems that prevent small businesses from expanding into the formal economy, while at the same time creating opportunities for corruption and graft.[32] Furman, Porter and Stern suggest that a nation's innovative capacity depends on the presence of "a strong common innovation infrastructure, cross-cutting factors contributing to innovativeness throughout the economy . . . public policy plays an important role."[33]

New companies formed with ease

New companies are formed with ease in America: registration is easy, franchise fees for SMEs are comparatively low, regulatory "paperwork" is minimal, and little if any bureaucratic oversight is required for daily corporate operation. Ease of market entry is critical to commercial success as rapid response to perceived opportunities—"time to market"—is often proven to be more important than other business considerations, including IP protection. As an example, in negotiation of an agreement with Microsoft for sponsored research at Texas A&M University several years ago, the Microsoft negotiator responded to my efforts to negotiate IP terms for the agreement by saying: "Microsoft does not care about intellectual property rights in research; simply give us the freedom to operate with the information gained from the research and we will either incorporate the knowledge into a product in six months, or we will have moved on to something new" (conversation between Microsoft Representative and author, *c*.2002). In the US, SMEs can respond to innovative business opportunities quickly, as ease of entry into the marketplace is simple.

In the last ten years, a tremendous facilitator to SME formation has been the advance of online commerce. For instance, it has been noted that if eBay employed the people who earn their income selling on its site, it would be the nation's second largest private employer, only behind Wal-Mart.[34] As an international example of the importance of ease of market entry to SME growth, the World Bank reports that because of new and easier regulations implemented in 2004, Serbia reduced the number of days to start a business from fifty-one days in 2003 to fifteen days in 2004, and the cost from 5,000 euros in 2003 to 500 euros in 2004. As a result, the number of new SMEs entering the market in Serbia leapt by 42 percent in 2004 alone, compared to 2003.[35]

Antitrust laws conducive to SME competition

Antitrust laws in the US—at both federal and state levels of government—insure competition and prevent corporate monopolies, assuring that SMEs have opportunities to enter new markets. Antitrust laws are extremely complex, too detailed for description in this chapter. However, the laws are strictly enforced, creating an atmosphere where SMEs indeed have opportunities in the face of monopolistic practices. For example, in the software industry, antitrust actions filed against Microsoft are frequently in the news. As another example, on December 19, 2005, the State of Alaska charged BP Plc and Exxon Mobile Corporation—the world's largest publicly traded oil companies—under antitrust laws with conspiring to withhold natural gas from US markets and reinforce their market monopoly over Alaskan supplies, preventing competition from other smaller companies.[36]

Taxation regime conducive to SME growth

Both personal taxes and corporate taxes in the US are transparent, predictable, and graduated (based upon income level, rather than "lump sum"). In the US, corporate taxes are levied by federal and state governments, and the rates vary from state to state; five states levy no tax whatsoever on corporate income. A 2005 study found that those regions of the US with the most "business friendly" tax rate have the highest rates of entrepreneurship (highest rate of new start-up SMEs per capita), and that "the greatest gains in entrepreneurship can be had by reducing government-imposed tax burdens on entrepreneurs and other businesses [SMEs]."[37] This "business-friendly" tax structure in the US is most conducive to the growth of innovation-based SMEs by stabilizing expectations, permitting strategic planning, and making taxes affordable as they are based upon income. In comparison, the World Bank has reported that national taxation structure is the greatest barrier to the growth of SMEs around the world (10,000 firms in 80 countries were studied).[38] Another recent study concluded that in the independent states of the former Soviet Union, changing the system of SME taxation would be the quickest way to create strong national SME market segments; the tax structure "has a positive direct effect and the additional indirect benefit of also lowering regulatory burdens and corruption. Former communist countries that have implemented such taxes have quickly created a large small-scale business sector."[39]

Tax credits for R&D support SME growth

Tax credits for corporations that make investments in "research and experimentation" or "research and development" in the US are equally complex, but also have provided great impetus for innovation development.[40] Corporations may offset their tax liability by a percentage of investments in R&D; some SMEs may even receive a cash sum in return for their R&D investments, both cases providing incentives to invest in R&D and promote innovation. A 2005 OECD study found that "tax reliefs for private R&D are found to provide a stronger stimulus [upon innovation], on average, than direct government subsidies."[41] The current federal law providing for tax credits terminated on December 31, 2005. However, a new bill has been introduced and is under debate in Congress that would extend the R&D tax credit for an additional four years, and would even increase the credit rate; if passed as proposed, the bill will result in a $9.8 billion annual benefit to all industries engaging in R&D, 2006–9, while dramatically increasing R&D investments.[42]

Availability and cost of reasonable credit

One of the most significant barriers to growth faced by high-tech SMEs internationally is the cost and availability of credit; in the US, credit is easier to obtain than in most countries of the world. First, the banking and equity markets are stable. Additionally, the government provides assistance through programs discussed previously (SBIR, STTR, ATP, and others) as well as through loan programs administered by the SBA. Specifically, the "Section 7(a)" loan program authorizes the SBA to guarantee bank loans up to $750,000 to SMEs that cannot otherwise obtain financing on reasonable terms, at a guaranteed rate of 75 percent of the total loan. Another SBA initiative—the Section 504 Program—provides direct long-term financing to SMEs to acquire or construct new facilities for their operations or to purchase machinery and equipment with a useful life of ten years or more.[43]

"Friendly" bankruptcy laws

In the creation and/or daily operation of an SME, failure is an ever-present possibility. Bankruptcy laws enable SMEs to close down an unsuccessful business. When properly crafted, bankruptcy laws can have a positive effect on the level of entrepreneurship; previously, it was demonstrated that the majority of those SMEs who fail in the US start again to form a new company. America has "gentle" bankruptcy laws assuring that the personal finances of small business persons are not ruined by corporate bankruptcy, unlike many nations of the world where personal and corporate finances are inseparably linked. Other countries are beginning to recognize that relaxed bankruptcy requirements spur economic development: Japan, Brazil, Finland, Indonesia, Portugal, and Vietnam passed new bankruptcy laws in 2004. The World Bank recently reported a "striking correlation" between efficient bankruptcy laws and the availability of bank and trade financing, stating that "bottlenecks in bankruptcy tend to be a strong deterrent to investment."[44]

Other supportive factors in a national innovation system

The preceding listing of five important SME-friendly initiatives created directly or indirectly by the US government is not all-inclusive. Other important factors beyond the scope of this paper include: (1) labor market flexibility (the OECD suggests that there is a strong negative relationship between the strength of employment protection legislation and R&D intensity);[45] (2) strong and stable equity markets; (3) laws supporting and enabling the ease of transfer of research results from the public sector to the private sector, such as the well-known Bayh-Dole Act of 1980, and similar acts related to research at the nation's federal laboratories; and (4) external financing that is available to SMEs (VC and FDI).

Other intangible factors contributing to America's success

Innovation does not occur in a vacuum; America's success in innovation and related economic development occurred in a land rich in national resources, a blessing that played a significant role in its success. As Dr Sachs observes in *The End of Poverty*:

> Americans, for example believe that they earned their wealth all by themselves. They forget that they inherited a vast continent rich in natural resources, with great soils and ample rainfall, immense navigable rivers, and thousands of miles of coastline with dozens of natural ports that provide a wonderful foundation for sea-based trade. Other countries are not quite so favored. Many of the world's poorest countries are severely hindered by high transport costs because they are landlocked; situated in high mountain ranges; or lack navigable rivers, long coastlines, or good natural harbors.[46]

Additionally, the size of consumer markets must be considered; comparisons of SME regional performance often are made without due consideration given to scales of markets. A Finnish associate suggested to me in Helsinki in September 2005:

> American small businesses are lucky with a distinct advantage over Finnish SMEs. When your companies open their doors, they have a potential market of 200 million people plus within reach, a tremendous opportunity! When we open our doors in Finland, we have only two meters of snow.[47]

Obviously, his point was that a national market of 200 million potential customers presents opportunities for SMEs that are "orders of magnitude" greater than those presented by the national population of five million Finns.

Additionally, the US presents a homogeneous market of immense proportions with no borders or language barriers. Mr Campbell Warden, editor of the book *Reflections on the Role of the Research Infrastructures in the European Research Area*, commented to me in response to a recent survey on reasons for SME failure: "Despite all the hype of a single market and other European Union literature, the reality is that linguistic barriers and geographical trading cultures mean that an SME in Europe faces severe limitations to its growth beyond its own national borders."[48] SMEs in the US do not face similar linguistic or cultural barriers to trade in the world's largest consumer market, a distinct advantage to SMEs in America.

These factors of geography, homogeneity, and market size cannot alone account for the gap in innovation recorded by the United States over other countries in the latter half of the twentieth century, as numerous other countries in the world have similar features. Dr Sachs provides an excellent "geopolitical" analysis of the development of nations after the Second World War, suggesting that each nation took one of three paths: (1) the countries already industrialized as of 1945 (Europe, the United States and Japan) reconstructed a new international trading system under US leadership, resulting in the General Agreement on Tariffs and Trade (GATT), the forerunner to today's World Trade Organization; (2) the socialist world remained cut off economically from the first group of countries until the fall of the Berlin Wall in 1989 and the end of the Soviet Union in 1991; and (3) the post-colonial countries basically stayed "nonaligned" from the other two groups, and were essentially isolated from global economic progress, innovation, and the advance of technology through to the end of the twentieth century.[49]

The industrialized countries that rebuilt the trading system after the Second World War were characterized by: political stability; rule of law, including clear and transparent government procurement regimes; equal enforcement of laws according to transparency and subject to the scrutiny and "freedom of the press"; lack of open bribery and corruption (while all societies—including America as amply demonstrated in the Enron fiasco—have corruption, these societies prosecute those who engage in corruption when identified). All of these factors have established a cultural and political environment in the US that nurtures innovators and makes possible an ecosystem where innovation is encouraged and rewarded.

Looking ahead—the twenty-first century

National borders and the geographical barriers to SME market success are falling in the commercial world, much akin to the "fall of the Berlin Wall" in the political world in November 1989. In *The World is Flat*, Dr Friedman observes:

> Clearly it is now possible for more people than ever to collaborate and compete in real time with more other people on more different kinds of work from more different corners of the planet than at any previous time in the history of the world—using computers, e-mail, networks, teleconferencing and dynamic new software . . . we are now connecting all the knowledge centers on the planet together into a single global network, which—if politics and terrorism do not get into the way—could usher in an amazing era of prosperity and innovation.[50]

No country or region of the world can "sit upon its laurels" as globalization is the order of the day. Furthermore, innovation—the driving force behind globalization—is changing the business environment so quickly that traditional SME practices that may have sufficed even ten years ago will not assure survival today in Europe, Japan, or the US. Despite creating the knowledge economy in the twentieth century, the US today is home only to six of the top twenty-five most competitive information technology companies in the world. An internationally accepted computer operating system or platform, the "digitizing" of every aspect of our lives, and the spread of the internet and the World Wide Web with its browser searching capabilities, have opened the world of information to all. Traditional barriers to SME growth are falling. Accordingly, *international networking* is required today for corporate survival, for companies large and small. Clearly, the rules of the business game are changing. SMEs must be flexible but, most important, internationally networked to play in this new game.

Notes

1 William A. Wulf "Review and renewal of the environment for innovation," unpublished paper, 2005, quoted in *Rising Above the Gathering Storm: Energizing and Employing America for a Brighter Economic Future*, Washington DC: The National Academy of Sciences, 2005, chapter 8, p. 1.
2 Jeffrey D. Sachs *The End of Poverty: Economic Possibilities for Our Time*, New York: Penguin Books, 2005, p. 41.
3 Pauline Maier, Merritt Roe Smith, Alexander Keyssar, and Daniel J. Kevles *Inventing America: A History of the United States*, New York: W. W. Norton & Company, Inc., 2002.
4 Sachs *The End of Poverty*, p. 61.
5 Jeffrey L. Furman, Michael E. Porter and Scott Stern "The determinants of national innovative capacity," *Research Policy*, 31, 2002, p. 900.
6 America's economic growth in the latter half of the twentieth century is not utopian, by any means. The United States is far from perfect, income gaps between the poor and rich in society are greater than similar gaps in the EU, and periodic cycles or "waves" of economic change have fundamentally reordered the organization of business and markets and even the role of government in the innovation process. Nearly 12 percent of the country's population live in households with less 40 percent of the median income, and there are more than two million children who have no health insurance of any kind. When people in a 2005 survey of sixteen countries were asked "What is the most attractive place in which to lead a good life?", respondents in only one country (India) indicated the United States. See *Rising Above the Gathering Storm: Energizing and Employing America for a Brighter Future*, Executive Summary, Washington DC: The National Academy of Sciences, 2005, p. ES-8.
7 My initial thoughts on the cultural business norms in America must be attributed to Dr Charles D. Wessner of the National Academy of Science, from an outstanding presentation he delivered to the Russian–US Innovation Council on High Technologies in Moscow in July 2005.
8 Mikel Landabaso "Reflections on US economic development policies: meeting the 'new economy' challenge," Visiting EU Scholar, Chapel Hill NC: University of North Carolina, p. 10. Unpublished paper at: http://.in3.dem.ist.utl.pt/inov2001/files/m_landabaso.pdf.

9   US SBA referenced in *Fiscal Notes*, Austin TX: John Sharp, Texas Comptroller of Public Accounts, September 1996.

10 Thomas L. Friedman *The World is Flat: A Brief History of the Twenty-first Century*, New York: Farrar, Straus and Girous, 2005, pp. 104–5.

11 *Losing the Competitive Advantage?* Executive Summary, Washington DC: American Electronics Association, 2005, p. 21.

12 *United States Constitution*, Article 1, Section 8.

13 The US Patent Code undergoes change with each Congressional session. Topics under current (and heated) debate include conformity of the US code with international standards and the impact of "open source" and software, as the nature of technology changes. Even the NII, created in 2004, acknowledged "[a] balanced legal regime that both protects the rewards of intellectual property and facilitates the spread of open standards is one of the requisites for an American Innovation Century." See *Innovate America: Thriving in a World of Challenge and Change*, Foreword, Washington DC: NII, 2004, p. 4.

14 Florence Jaunotte and Nigel Pain *From Ideas to Development: The Determinants of R&D and Patenting*, Paris: OECD, Economics Department Working Paper No. 457, 2005, p. 2.

15 Florence Jaunotte and Nigel Pain *Innovation in the Business Sector*, Paris: OECD, Economics Department Working Paper No. 459, 2005, p. 37.

16 The $284 billion invested in R&D in the US was larger than the R&D expenditures for Japan and the EU combined for the same time period. As a percentage of GDP, the expenditure represented 2.65 percent of GDP, below Japan's 3.06 percent investment, but above most other industrialized countries of the world. See *AAAS Report XXIX: Research and Development FY 2005*, Chapter 1, Washington DC: American Association for the Advancement of Science, 2005. Accessed at www.aaas.org/spp/rd/05pch1.htm.

17 *Losing the Competitive Advantage?* Executive Summary, Washington DC: American Electronics Association, 2005, p. 14.

18 Jeffrey Mervis "US science policy: bill seeks billions to bolster research," *Science*, Vol. 310, No. 5756, p. 1891. Emphasis added.

19 *AUTM US Licensing Survey: FY 2004*, Survey Summary, Northbrook IL: AUTM, 2005, pp. 5–6.

20 Ross DeVol and Perry Wong *America's High-tech Economy: Growth, Development and Risks for Metropolitan Areas*, Los Angeles: Milken Institute, 1999.

21 *AAAS Report XXIX: Research and Development FY 2005*, Chapter 1, Washington DC: Milken Institute, American Association for the Advancement of Science, 2005. Accessed at www.aaas.org/spp/rd/05pch1.htm.

22 See www.sba.gov/SBIR/indexsbir-sttr.html#sbir.

23 See www.sba.gov/SBIR/indexsbir-sttr.html#sbir.

24 See www.atp.nist.gov/atp/overview.htm.

25 See www.mep.nist.gov/.

26 See www.nsf.gov/div/index.jsp?div=EPSCOR.

27 See www.sba.gov/sbdc/.

28 See www.osec.doc.gov/%3Fbmi%3F/budget/05BIB/eda.pdf.

29 See www.nsf.gov/publications/pub_summ.jsp?ods_key=nsf01116.

30 Landabaso "Reflections on US economic development policies," p. 4. Emphasis added.

31 William Schweke *Understanding State Business Climates*, Washington DC: Corporation for Enterprise Development, 2000, p. 18.

32 World Bank *Doing Business in 2006*, Washington DC: World Bank, 2005.

33 Furman, Porter and Stern "The determinants of national innovative capacity," pp. 900 and 905.

34 See www.naplesnews.com.
35 World Bank "World Bank Press Review," Washington DC: World Bank, September 17, 2005.
36 John R. Wilke and Russell Gold "BP, Exon hit with antitrust suit," *Wall Street Journal*, December 20, 2005, p. A-3.
37 Thomas A. Garrett and Howard J. Wall *Creating a Policy Environment for Entrepreneurs: Working Paper 2005–064A*, St Louis MO: The Federal Reserve Bank of St Louis, 2005, pp. 8 and 18.
38 World Bank *World Business Environment Survey*, 2000. Accessed at www.ifc.org/ifcext/economics.nsf/Content/ic-wbes.
39 Anders Aslund and Simon Johnson *Small Enterprises and Economic Policy*, Washington DC: Carnegie Endowment for International Peace, 2003, p. 2.
40 For an excellent summary of US R&D tax credits, see Bronwyn H. Hall *Tax Incentives for Innovation in the United States: A Report to the European Union*, Oxford: Oxford University, 2001.
41 Jaunotte and Pain *Innovation in the Business Sector*, p. 9.
42 Roxanna Tiron "Defense lobbyists try to shepherd long-sought R&D tax break through house," *The Hill*, January 2, 2006. Accessed at www.hillnews.com/thehill/export/TheHill/Business/113005_tax.html.
43 Current information on both the SBA Section 7(a) and Section 504 loan programs can be found at: www.sba.gov/financing/index.html.
44 World Bank *Doing Business in 2006*, Washington DC: World Bank, 2005.
45 Florence Jaunotte and Nigel Pain *An Overview of Public Policies to Support Innovation*, Paris: OECD, Working Paper 456, 2005, p. 23.
46 Sachs *The End of Poverty*, p. 57.
47 Conversation with Janne Virtapohja, Foundation for Finnish Inventions, September 16, 2005.
48 Campbell Warden in Terry A. Young (ed.) *Failure of University Spin-off Companies*, 2005. Accessed at www.terryyoungllc.com/publications_and_resources.html.
49 Sachs *The End of Poverty*, pp. 46–9.
50 Friedman *The World is Flat*, p. 8.

6 Cutting the costs of innovation R&D through business partnering in Japan

Takuma Kiso

Although economists nowadays claim that the Japanese economy has finally hit bottom and is now on a gradual rise, views among business owners have been divided. This reflects the fact that past economic recoveries were felt and shared by many firms while the current economic upturn can be experienced by only a section of businesses. One of the main reasons that divide businesses into the winning and losing groups is whether a company succeeds in or fails in constant innovation. Innovation, however, is very costly.

This chapter analyzes some apposite cases of business partnering for the purpose of doing research for innovation, which is an urgent requirement for current corporate management. First, M&A and tie-ups among big businesses in two industries in Japan, pharmaceutical and consumer electronics industries, are discussed. Case studies include the start-up of Astellas Pharma Inc. by the merger of Yamanouchi Pharmaceutical Co., Ltd and Fujisawa Pharmaceutical Co., Ltd and the establishment of the TV liquid crystal display (LCD) panel joint venture IPS Alpha Technology Ltd by Hitachi Ltd, Toshiba Corporation, and Matsushita Electric Industrial Co., Ltd (MEI). Then a new type of business partnering among SMEs is introduced, and industry–university cooperation is also discussed in this chapter.

Partnering among large companies in Japan

In the 1990s and after, the Japanese economy lost momentum and various industries and companies had to go through a restructuring process. One of the most popular ways of restructuring was to use M&A, while partnering has also been used. This section looks at M&A and partnering in the pharmaceutical and consumer electronics industries in Japan.

Pharmaceutical industry

In and after the year 2000, the pharmaceutical industry has seen many M&A. The reasons behind this phenomenon will be made clear by a detailed examination of Astellas Pharma Inc.

In April 2005, Astellas Pharma Inc. was created through a merger of Yamanouchi Pharmaceutical Co., Ltd (ranked third in the Japanese market) and Fujisawa Pharmaceutical Co., Ltd (ranked fifth in the Japanese market). The background and objectives of the merger are described in a news release dated February 24, 2004. The first paragraph explains how the two companies see their business environment. It should be noted that, in the globally competitive pharmaceutical industry, the importance of large but efficient R&D is emphasized:

> With the growing pressure to control medical expenses in major developed countries, intensifying global competition for the development of new drugs and rising R&D spending, the business environment surrounding the pharmaceutical industry has been increasingly challenging. Competition in the domestic market has intensified as well, characterized by the further implementation of medication cost control policies, such as drug price cuts, and market penetration by global pharmaceutical companies. Under these circumstances, in order to compete globally and achieve sustainable growth, it is necessary to spend on R&D to create innovative new drugs, as well as to develop a global platform to recover efficiently the ever-increasing investment costs.[1]

They go on to declare their decision to merge:

> Based on such common understanding by Yamanouchi and Fujisawa, the companies have been considering a merger to enhance their core business platform, the ethical pharmaceutical business and to succeed in the global arena where competition is ever-intensifying. Today, Yamanouchi and Fujisawa reached an agreement on the basic terms of their merger.[2]

The third paragraph describes an image of the new company with its main objectives:

> The Combined Company aims to achieve economies of scales through the integration of the R&D, and sales and marketing capabilities of the both companies, as well as to improve further its profitability through the establishment of more efficient operational structures. The Combined Company will stand as a completely new entity, which is neither Yamanouchi nor Fujisawa, and will strive to contribute to health of the people around the world as a global pharmaceutical company with excellent R&D, and sales and marketing capabilities of its own.[3]

This is not the only case of M&A in the Japanese pharmaceutical industry. In 2001, for example, Mitsubishi Pharma Corporation, which belongs to the third tier in the industry, was established by a merger of Welfide Corporation and Mitsubishi-Tokyo Pharmaceuticals Inc. Welfide Corporation was established by a merger of Yoshitomi Pharmaceutical Industries, Ltd and Green Cross Corporation in 1998, and Mitsubishi-Tokyo Pharmaceuticals Inc. was established by a merger of Tokyo Tanabe Co., Ltd and Mitsubishi Chemical Corporation in 1999. And October 2005 saw three M&A: (1) the creation of Dainippon Sumitomo Pharma in October 2005 through a merger of Dainippon Pharmaceutical Co., Ltd and Sumitomo Pharmaceuticals Co., Ltd, both of which belonged to the fourth tier in the industry; (2) the establishment of Daiichi Sankyo Company, Ltd as a stock holding company of Dainippon Pharmaceutical Co., Ltd and Sankyo Co. Ltd; and (3) the formation of Aska Pharmaceutical Co., Ltd by a merger of Teikoku Hormone Manufacturing Co., Ltd and Grelan.

As seen above, M&A have been very much accelerated in the last few years. Although some of them failed shortly after their announced intentions, most of them succeeded at least superficially. Why have so many cases been arising?

First, it can be pointed out that circumstances surrounding the Japanese pharmaceutical industry have been very tough. In the global market, as a result of various M&A, mega-pharmaceutical companies have been born. It is said that the critical mass for a drug firm that can survive the global market has been raised from $10 billion a few years ago to $20–30 billion today.[4]

This reflects the fact that the average costs of developing and marketing a completely new drug that will be accepted in the global market have become astronomical. The costs are said to have increased from $100 million ten years ago to $500–800 million today. In other words, in order to place two or three new drugs on the market every year, between $1 billion and $2 billion should be allocated to R&D annually.

Second, the high costs of developing new drugs increased the cases of M&A. Why, then, have the costs become so high? Because the development of a new drug requires several time-consuming processes. According to the website of Shin Nippon Biomedical Laboratory Ltd,[5] five stages are needed for a new medicine to be marketed: (1) basic research for two to three years, including research for new compounds and physical chemistry research; (2) pre-clinical studies for three to five years, including safety studies, safety pharmacology studies, efficacy pharmacology studies, and pharmacokinetic studies; (3) clinical studies for three to seven years, including a phase I trial (clinical pharmacology trial), a phase II trial (exploratory trial), and a phase III trial (confirmatory trial); (4) drug approval for two to three years, including new drug application, investigation, and drug approval; and (5) post-market studies, including manufacturing/marketing and post-marketing surveillance.

Third, the fact that patents of some medicines will expire by 2010[6] will press drug makers to strengthen and hasten their R&D activities. The Japanese patent of Sankyo's anti-hyperlipidemic drug "Mevalotin," for instance, expired in 2002 and the patent for the US will expire in 2006. The US patent of Fujisawa's immunosuppressant "Prograf" will expire in 2008, that of Takeda's anti-ulcer drug "Takepron" in 2009, Eisai's Aricept for treatment of severe Alzheimer's disease in 2010. In order to keep high profits, pharmaceutical companies need candidate medicines (medicines in the pipeline) at the various stages of R&D.

Fourth, in the domestic market, ongoing deregulation of medicine and the expanding presence of foreign drug firms have been pressuring Japan's drug firms to restructure themselves. The national health insurance drug prices have been revised downward. Use of cheaper generic drugs has been recommended by the government. Marketing powers of foreign drug firms have been increased against the background of M&A occurring among themselves.

While large drug firms have been coping with these increasingly tougher environments through M&A, relatively small pharmaceutical companies have chosen partnerships and tie-ups. Some of the most recent examples are as follows.

In July 2005, Taisho Pharmaceutical Co., Ltd, which belongs to the third tier in the industry, announced that it would forge business and capital ties with Yomeishu Seizo Co., Ltd. Their news release dated July 11, 2005 set out the purpose of the partnership. Like the merger of Yamanouchi and Fujisawa, the first paragraph shows how Taisho and Yomeishu Seizo see their business environment. It should be noted that Taisho and Yomeishu Seizo pay attention to the domestic market while Yamanouchi and Fujisawa are more interested in the global market:

> Health-consciousness among consumers has continued to rise in recent years, while the growing trend toward self-medication has led to increasingly diverse consumer needs. Taisho's operating environment, meanwhile, is undergoing unprecedented changes amid a contraction in the over-the-counter medications market and structural changes in distribution channels sparked by deregulation.[7]

They carry on to declare that the two companies will make a partnership:

> In this context, Taisho has chosen to pursue a business partnership with Yomeishu Seizo to respond to the broadening range of products expected to be available on the self-medication market. This decision also reflects Taisho's desire to prepare for competition that transcends the bounds of its traditional medications and food products, as well as the necessity to devise initiatives driven by new ideas and innovative perspectives for growth opportunities in health-related markets.[8]

This then leads to a specific form of business partnership, namely a capital tie-up:

> By taking advantage of Yomeishu Seizo's expertise in natural medicines, Taisho is seeking to develop new products and create new markets. The partnership agreement includes a capital tie-up to enhance the overall effectiveness and strength of the partnership.[9]

In the same month, Kobayashi Pharmaceutical Co., Ltd announced that it had recently concluded a contract for capital tie-up operations with Itoh Pharmaceutical Co., Ltd. Their objective is explained in their news release dated July 1, 2005, and the following quote explains what Kobayashi has been achieving in recent years:

> In its household product manufacturing and sales business, the Company has positioned as its focus category general purpose pharmaceuticals, health food and oral hygiene products, has aggressively committed management resource to them, and expanded its product line-up to accommodate wide-ranging and diversified consumer needs and the demands of an aging society.[10]

They then refer to what the partner, Itoh, has been doing:

> Itoh, meanwhile, with in-house development and manufacturing and outsourced manufacturing as its main supports, has fully shown its competitively superior manufacturing and planning and development capabilities, and strongly expanded its business content.[11]

Finally, they explain the aims of the capital tie-up of the two companies:

> In these circumstances, it is adjudged that interactive sharing of product development, technological and sales capabilities, as well as management expertise, would enable expectations for further development of both companies and improvement of corporate values, leading to the decision for a capital tie-up.[12]

In September 2005, Rohto Pharmaceutical Co., Ltd, which belongs to the fifth tier in the industry, announced that it had agreed with Pharma Foods International Co., Ltd joint developments and capital alliance. By this agreement, Rohto Pharmaceutical Co., Ltd, which has two major business lines – eye care products and skin care products – will be able to strengthen its third core business line, health-enhancing food and supplements.

As seen above, the Japan's pharmaceutical industry has been changing its structure through M&A among large drug makers and business tie-ups among medium-sized ones in order to make their R&D activities more efficient so that they can survive the current, radically changing business environment.

Consumer electric industry

Japanese electric appliance makers have had overall business lines from white goods to information appliances to industrial electric appliances. In the prolonged recession of the Japanese economy, however, each electronic manufacturer has focused on its strongest business lines. During this process partnering and tie-ups are often used. However, whole companies do not partner or tie-up: a division of a company partners or ties up with a counterpart of another company. In other words, company A ties up, or sets up another firm, with company B for its production of X while company A establishes another company with company C for its production of Y. At this juncture, it is appropriate to review specific cases in the production of semiconductors, flat-screen televisions, and mobile phone handsets.

One of the most symbolical businesses for tie-ups among consumer electric companies is the manufacturing of one type of semiconductor, dynamic random access memory (DRAM). Towards the end of the 1990s, five Japanese mega-electric appliance makers, namely Toshiba Corporation, NEC Corporation, Hitachi Ltd, Mitsubishi Electric Corporation, and Fujitsu Ltd manufactured DRAMs. In November 1999, however, NEC and Hitachi announced that they had "signed an agreement to form a joint venture DRAM company at the end of December 1999 to be named 'NEC-Hitachi Memory Inc.' Operations were slated to begin from April 2000."[13] Both NEC and Hitachi are shareholders of the company but are not involved in manufacturing. The new name "Elpida Memory Inc." was given to the company in September 2000. Then in April 2003 Elpida took on Mitsubishi Electric's DRAM operations while it formed a new partnership with Powerchip.

In June 2002, Fujitsu and Toshiba "announced that they had agreed to explore a comprehensive collaboration focusing on system-on-chip (SoC) solutions … Aiming to provide SoC solutions at 100 nanometers and finer, the companies" have established "several joint working groups to investigate collaboration in such areas as the standardization of design and development platforms and silicon technology, co-development of processor cores and other intellectual property, and the joint development of advanced LSIs for communications and other fields."[14]

In April 2003, Hitachi and Mitsubishi Electric established Renesas Technology Corp., which is engaged in the "development, design, manufacture, sales and servicing of system LSIs (large-scale integrations), including microcomputers, logic and analog devices, discrete devices and memory products, including flash memory and SRAM (static random access memory)."[15]

Another interesting example of partnering among electric companies is seen in the production of flat-screen televisions. There are three technologies for flat-screen televisions, and companies in each technology camp have been co-operating. In terms of plasma TVs, Hitachi Ltd and MEI announced, on February 7, 2005, that they would "be joining forces to develop and expand the plasma TV market under an agreement that calls for comprehensive collaboration in moving forward with the plasma display panel (PDP) businesses."[16] Five days before this news, Fujitsu Ltd had announced that it had "reached a basic agreement with Hitachi Ltd regarding Fujitsu's transfer to Hitachi of 30.1 percent of the issued shares in their joint venture company, Fujitsu Hitachi Plasma Display Limited (FHP), along with related plasma display panel (PDP) intellectual property rights."[17]

As for LCDs, Fujitsu Ltd announced on February 5, 2005 that it had "entered into a basic agreement with Sharp Corporation regarding the transfer of Fujitsu's liquid crystal display (LCD) operations to Sharp."[18] This followed the announcement on October 29, 2004 by Hitachi Ltd, Toshiba Corporation, MEI, and wholly owned Hitachi subsidiary Hitachi Displays Ltd that they "officially concluded a joint venture agreement to establish a company to manufacture and sell LCD panels for flat panel TVs."[19] The joint venture IPS Alpha Technology Ltd commenced its operations on January 1, 2005. Before this, on March 24, 2004, Sanyo Electric Co., Ltd and Seiko Epson Corporation announced that "they would merge their liquid crystal businesses to form a new company. The new company was known as Sanyo Epson Imaging Devices Corporation and planned to begin operations in October 2004. Epson was to hold 55 percent of the joint venture, and SANYO 45 percent."[20]

On July 15, 2004, Sony Corporation and Samsung Electronics Co., Ltd announced the grand opening of their new building for manufacturing S-LCD Amorphous thin film transistor (TFT) LCD: "S-LCD corporation was a joint venture company that was established on April 26, 2004. It was an equal joint venture by Samsung Electronics Co., Ltd (hereafter Samsung) and Sony Corporation (hereafter Sony) dedicated to the manufacturing of amorphous TFT LCD panels for LCD TVs."[21]

A third example of partnering lies in the manufacturing of mobile phone handsets:

In recent years, handset manufacturers have been steadily moving toward the implementation of two main OS standards as the advanced mobile operating system for FOMA handsets: Symbian OS, which has a strong track record overseas as an operating system for mobile phones, and Linux, which is well-known as an open standard operating system.[22]

The alliance of Sharp and Sony Ericsson Mobile Communications Japan and that of Fujitsu and Mitsubishi Electric chose Symbian OS while that of NEC and MEI chose Linux. Specific tie-up cases follow.

On November 29, 2004, Sharp Corporation and Sony Ericsson Mobile Communications Japan, Inc. "jointly announced an agreement to co-develop the base software and to share selected hardware for 3G FOMA (freedom of multimedia access) mobile phones for NTT DoCoMo, Inc."[23] On March 24, 2004, Fujitsu Limited and Mitsubishi Electric Corporation announced that "they are exploring a collaboration to jointly develop new FOMA mobile handsets running the Symbian OS operating system for NTT DoCoMo, Inc."[24] Before these cases, on August 21, 2001, NEC Corporation, MEI, and one of MEI's principal subsidiaries, Matsushita Communication Industrial Co., Ltd, announced that "they have signed an agreement to form a development alliance aimed at strengthening global business expansion of their technologies and products in the field of advanced mobile handsets."[25]

Partnering among SMEs in Japan

Partnering, however, has not been an exclusive feature of large-sized corporations. It has also been a phenomenon commonly seen among SMEs.

Partnering among SMEs has not been a rare thing

Partnering among SMEs is not a new phenomenon in Japan. It is often said that one of the main characteristics of the Japanese SMEs is that they have been subcontractors to large manufacturers. They are at the bottom of Japan's industrial pyramids. This, however, is only a part of the SMEs story. SMEs have had a horizontal network as well. According to a book entitled *Machi-Koba Sekai wo Koeru Gijutu Hokoku (Report on Town Workshops' Technologies Exceeding the World Standards)* written by Tomohiro Koseki, a back-street factory that makes metal cutting machine tools cooperates with 100 workshops as its suppliers and subcontractors.[26] They were located not only in Tokyo but also in Saitama, Kanagawa, Aichi, and Osaka Prefectures, and in Sweden, Germany, and the US. And their businesses varied from casting to can making, heat treatment, boring, planing, gear cutting, polishing, electric wiring, and noise and vibration insulation.

Also, a phrase "cross-industry association" has been one of the key phrases when developments of SMEs are discussed. But in reality many associations have turned out to be pan-industry social event groups and/or pan-industry study groups without manufacturing any products. In other words, these groups often function only as networking groups for the purpose of socializing and/or lecture meetings that invite various specialists as their lecturers. In the better cases, they have been mutually cooperative societies that share their workloads at the time of both booms and busts.

What is new to the recent partnering among SMEs? The case of Rodan21

New types of cross-industry associations, which aim for partnering, have been appearing in recent years. One of the most active and innovating partnering examples is Rodan21. It is a network of SMEs that behave like a corporation. It will therefore be very useful to look at this group in detail in order to exemplify the nature of some of the most needed and successful SME models.

According to the website of Creation Core Higashi Osaka, Rodan21 Inc. is

> a group of small and medium-sized enterprises centered on manufacturers based in Higashi-Osaka City and established to support companies in their manufacturing activities by means of a cross-industry network that provides planning, design, marketing and other services.
> [. . .] In carefully and rapidly responding to the diverse requests we receive from companies all over Japan, we are expanding our scope of activities by making maximum use of our manufacturing business network as well as our links to other enterprises, business groups and industrial support organizations nationwide and overseas enterprises.[27]

As can be seen from this statement, Rodan21 Inc. is now a unique partnering system with a form of Kabushiki Kaisha (limited company). Its origin, however, was a pan-industry study group. The same website tells the history of the company: "Rodan21 was set up by Higashi-Osaka City Office as a new form of government-established, privately-run group that is seeking to move away from its public sector roots by expanding its original network."[28]

According to a book entitled *Higashi-Osaka Genki Kojo* (*Vigorous Workshops in East-Osaka*) written by Takayuki Shinagawa, President of Rodan21, the company was established as a cross-industrial association group in 1997.[29] In that year, Higashi-Osaka City Office publicly sought companies who would like to participate in its cross-industrial integration program and twenty-five companies applied for this scheme. Somehow the applicants were divided into two groups, A and B. The former consisted of thirteen companies and began to call itself Rodan21 in April 1998. Then a year later the group set up a Yugen Kaisha (a limited liability company similar to an S corporation[30] in the US) and in May 2005 it became a Kabushiki Kaisha (a joint-stock company, or standard corporation), Rodan21 Inc.

Now the company is managed by fifteen companies and one secretariat. Instead of the names of companies, the lines of business of each company are used:[31] (1) plastic and rubber packing factory; (2) insulating rubber and rubber packing factory; (3) industrial cutter factory; (4) screws factory; (5) factory manufacturing arrows for darts and portable ashtrays; (6) cutting and drilling specialist; (7) precision sheet metal processing specialist; (8) fashion bags factory; (9) high-precision processing of plastic specialist; (10) industrial waste contractor; (11) engineering consultant; (12) plastic goods molding specialist; (13) factory manufacturing cork goods, parts of bicycles, and parts of shoes; (14) factory making educational materials and kits for art classes; (15) labor and social insurances specialist; (16) secretariat of Rodan21 and manufacturing coordinator.

The website continues by explaining how "[a]cting as a front desk for its members, we operate a one-stop service that accepts requests for idea development assistance from customer companies."[32] Here the meaning of the phrase "one-stop service" should be elaborated since it is one of the key phrases that characterize this new form of partnering among SMEs. Mr Shinagawa, President of Rodan21, emphasizes in his book *Higashi-Osaka Genki Kojo* that there is a great difference between a manufacturer and a back-street factory/workshop.[33] The former eliminates its production plans from the viewpoints of marketability and moves to test production of those passing the elimination processes. Then the commercialization process begins. This includes naming, packaging, and cataloging, and finally the products are marketed. On the other hand, the latter thinks of itself as a specialist of parts and lacks the viewpoint of the consumer.

Noticing this gap between a manufacturer and a back-street factory/ workshop, Mr Shinagawa has made Rodan21 as a coordinating company that provides services from production to the commercialization of articles by using a network of factories, designers, sales agents, trading companies, and licensed tax accountants. In other words, Rodan21 not only introduces relevant supplies, designers, distributors, and other related companies but also coordinates the production and commercialization process to companies who introduce ideas for products. The website says:

> Our system, which is supporting manufacturing from both the hardware and software sides by establishing the Rodan Total Institute (RRD: Rodan Research and Development), is capable of handling all the business details that accompany manufacturing and of conducting initial trials in the manufacturing field.[34]

The following paragraphs and the diagrams Figures 6.1 and 6.2 describe how Rodan21 conducts its businesses. They are quoted from the website.[35]

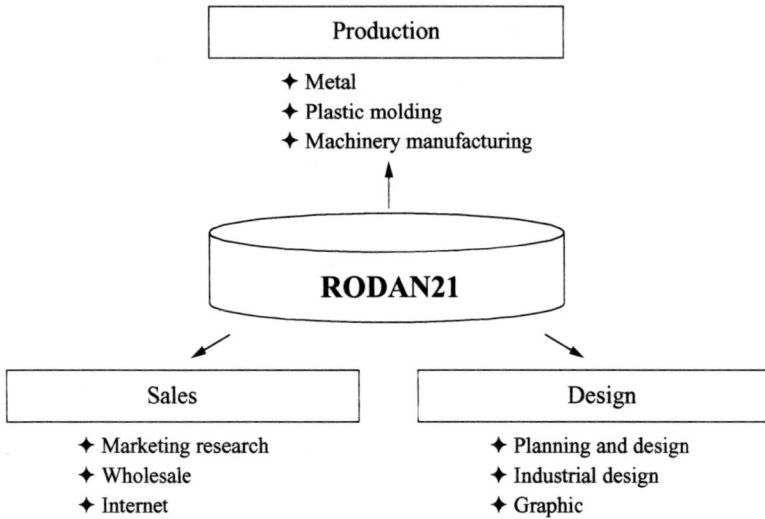

```
                    ┌─────────────────────────────┐
                    │        Production           │
                    └─────────────────────────────┘
                         ✦ Metal
                         ✦ Plastic molding
                         ✦ Machinery manufacturing
                                  ▲
                                  │

                    ╭─────────────────────────────╮
                    │          RODAN21            │
                    ╰─────────────────────────────╯
                    ↙                          ↘

  ┌──────────────────────┐        ┌──────────────────────┐
  │        Sales         │        │        Design        │
  └──────────────────────┘        └──────────────────────┘
     ✦ Marketing research            ✦ Planning and design
     ✦ Wholesale                     ✦ Industrial design
     ✦ Internet                      ✦ Graphic
```

Figure 6.1 Rodan21 as a collaborator for professionals of manufacturing, design and marketing

Source: www.m-osaka.com/en/exhibitors/024/products.html.

Collaboration

Rodan21's Product Development Supports Manufacturing in Collaboration with a Team of Professionals.

Our member clusters, which are active in a wide range of fields, and its members, who number approximately 140 people (limited to people who have signed confidential agreements), participate in the Planning & Development Committee and conduct discussions in which they express their opinions concerning different fields. In addition, it organizes support systems that answer the development needs of specific products, and it also performs manufacturing.

Support system

Rodan21 states that its unique product development system is centered on planning and development meetings. As seen in the following quotation, the company first analyses its clients' requests and then categorizes them in the meetings. Rodan21 makes

judgements concerning the contents of consultation cases based on the activities of Rodan21 member clusters, which are active in a wide range of fields. It invites prospective customers to talk to it about their requests and places the details of their proposals in order.

Then requests will be further examined, depending on their categories and its production phases. More specifically:

> In the case of proposals sent to the Planning and Development Committee, work proceeds while discussions are held and reports are issued on a weekly basis. It carries out coordination by adding necessary members in accordance with the proposal's state of progress, such as, for example, when cases need to be discussed by the Marketing Committee, or when design progress requires the case to be taken up by the Manufacturing Committee, etc.
>
> Moreover, [Rodan21] construct[s] a manufacturing process that takes each product from the trial production stage right through to commercialization.

Rodan21 also has established an academic–industrial alliance.[36] Rodan21 developed a new demand forecasting program with Shisutemu Sogo Kenkyuusho (Japanese Institute of System Research) and Higashiosaka Junior College. This program has been developed against the background that individual SMEs cannot afford to analyze the Japanese economy and to make demand forecast by themselves. This program can help them to simulate their sales and profits under such macro-economic shocks as the sharp appreciation of the yen.

Figure 6.2 Flow chart of commercialization by Rodan21

Source: www.m-osaka.com/en/exhibitors/024/products.html.

Mr Shinagawa, President of Rodan21, says in his book that in the first four years since the establishment of the company, only about 10 percent of the proposals brought to the company were successfully merchandized. With more sponsors and funds, success rates could be raised to 30 percent from the current 10 percent.[37] Successful merchandise includes a pump for fire extinguishing that can work under low levels of water, an anti-twisting device for domestic wiring, a unique armrest for personal computers, and a poncho that can be folded into a pocket-sized tissue pack. The challenges of Rodan21 are often featured in the media and have been encouraging owners of SMEs to think seriously about partnering.

Current Japanese government policy

The Japanese government, after seeing some successful examples, has understood that new types of partnering are one of the most important and effective ways to revive, revitalize, and strengthen Japanese companies, especially SMEs. This is reflected in the fact that the Ministry of Economy, Trade, and Industry (METI) has recognized that SMEs have only a limited amount of managerial resources and therefore that it will be a good idea for SMEs to supplement one another by making networks together and with research institutes.

Therefore, in April 2005, the three laws that had been supporting start-ups and business innovation—the Law on Supporting Business Innovation of Small and Medium Enterprises, the Temporary Law Concerning Measures for the Promotion of Creative Business Activities of Small and Medium Enterprises, and the Law for Facilitating the Creation of New Business—were integrated into a new law called the Law for Facilitating New Business Activities of Small and Medium Enterprises, and the aforementioned measures were made more robust.

According to the 2005 White Paper on Small and Medium Enterprises in Japan,[38] this law includes such measures as: (1) the promotion of start-ups; (2) business innovation support; (3) support for new partnerships; (4) development of conditions in other ways; and (5) other related measures. Among them the supporting measures for new partnerships are new. The White Paper clearly explains that "new business activities undertaken through organized collaboration between businesses in different fields and flexible collaboration effectively incorporating their business resources ('new partnership') is, in fact, a new objective." Some of the specific measures included are as follows. First

> the Government will certify and financially support "cross-field partnership new business development plans" formulated by collaborative groups incorporating elements such as the "reliable ascertainment of markets needs," "possibility of realization through mutual complementary relations and collaboration," "existence of core enterprises

capable of being the locus of external responsibility," and "existence of arrangements regarding e.g. process control and quality preservation."[39]

Second

local strategy councils to support new partnership will be established in each block, potential partners will be introduced and support for the operations and commercialization of partnerships and market development, etc. provided to businesses planning to engage in collaborative business, and diverse support up to commercialization stage will be provided from the business point of view.[40]

Third

support will be provided through advanced loans (20-year loan period, interest free) for the installation of joint facilities required for business undertaken by voluntary groups with approved plans.[41]

Other measures include capital investment tax reduction and preferential lending arrangements of government-affiliated financial institutions.

Concluding remarks

As seen above, partnering is a very common phenomenon among companies in Japan. For large companies, partnering is only one of the many options for corporate restructuring. Some have chosen M&A in pursuit of economy of scale while others have chosen partnering. In the case of Japan's pharmaceutical industry, top drug manufacturers have tended to chose M&A while smaller ones have tended to choose partnering. On the other hand, in the case of the consumer electric industry, many companies have chosen division-to-division partnering rather than company-to-company partnering.

For SMEs, partnering has been one of the few major options for their restructuring but its role has been changing recently. The old type of partnering has existed for load sharing while the new type of partnering is to form a virtual large company. Each participant in partnering behaves as if it were a division of one large corporation as seen in Rodan21. And the government has been encouraging SMEs all over Japan to establish this new type of virtual corporation.

Recent developments of new types of partnering, especially among Japan's SMEs, will surely lead the country's macro-economy on to a steadily growing path. Japanese companies still have the flexibility and robustness to cope with radical structural changes in the business environment.

Notes

1 Yamanouchi Pharmaceutical Co., Ltd and Fujisawa Pharmaceutical Co., Ltd "Yamanouchi and Fujisawa enter into a basic agreement to merge on April 1, 2005," news release, February 24, 2005.
2 Ibid.
3 Ibid.
4 "Large European and American pharmaceutical firms leading restructuring of large drug makers," *Syukan Toyo Keizai* (*Weekly Oriental Economy*), January 17, 2004, pp. 50–3.
5 See www.snbl.com/en/5000.html.
6 *Nikkei Sangyo Shimbun* (*Nikkei Business Daily*), February 13, 2005, p. 6.
7 Taisho Pharmaceutical Co., Ltd "Taisho forges business and capital ties with Yomeishu Seizo Co., Ltd," news release, July 11, 2005.
8 Ibid.
9 Ibid.
10 Kobayashi Pharmaceutical Co., Ltd "Notice regarding business tie-up contract," news release, July 1, 2005.
11 Ibid.
12 Ibid.
13 NEC Corporation "NEC and Hitachi establish joint venture DRAM company," press release, November 29, 1999.
14 Toshiba Corporation "Fujitsu and Toshiba to explore comprehensive collaboration in system-on-chip business," press release, June 19, 2002.
15 See www.renesas.com/fmwk.jsp?cnt=corporate_profile.htm&fp=/company_info/child_folder/&title=Company%20Infomation.
16 Hitachi Ltd "Hitachi and Panasonic agree on comprehensive collaboration in plasma display business," news release, February 7, 2005.
17 Fujitsu Ltd "Fujitsu announces agreement with Hitachi on plasma display panel business," press release, February 2, 2005.
18 Fujitsu Ltd "Sharp and Fujitsu announce agreement on transfer of Fujitsu's LCD business," press release, February 5, 2005.
19 News release of Hitachi Ltd dated October 29, 2004 entitled "Hitachi, Toshiba and Matsushita Conclude Agreement for Establishment of TV LCD Panel Joint Venture, IPS Alpha Technology."
20 Sanyo Electric Co., Ltd "Epson and SANYO liquid crystal businesses," news release, March 24, 2004.
21 Sony Corporation "Grand opening ceremony of S-LCD," press release, July 15, 2004.
22 Fujitsu Ltd "Fujitsu and Mitsubishi Electric explore collaboration on FOMA handset development for NTT Docomo," press release, March 24, 2004.
23 Sony Ericsson Mobile Communications Japan, Inc. "Sharp and Sony Ericsson announce co-development for 3G FOMA mobile phones in Japan," press release, November 29, 2004.
24 Fujitsu Ltd "Fujitsu and Mitsubishi Electric Explore Collaboration on FOMA Handset Development for NTT DoCoMo," press release, March 24, 2004.
25 NEC Corporation "Matsushita (Panasonic) and NEC agree to strategic cooperation in mobile handset business," press release, August 21, 2001.
26 Tomohiro Koseki *Machi-Koba Sekai wo Koeru Gijutu Hokoku* (*Report on Town Workshops' Technologies Exceeding the World Standards*), Tokyo: Shogakukan Inc., p. 31.
27 See www.m-osaka.com/en/exhibitors/024/index.html.
28 Ibid.

29 Takayuki Shinagawa *Higashi-Osaka Genki Kojo* (*Vigorous Workshops in East-Osaka*), Tokyo: Shogakukan Inc., 2003, pp. 17–18, 157.
30 An S corporation is "a form of corporation that meets IRS [Internal Revenue Services] requirements to be taxed under Subchapter S of the Internal Revenue Code [of the USA]" (www.en.wikipedia.org/wiki/S-corporation).
31 See www.rodan21.com/kaiin/rodan_kaiin.html.
32 See www.m-osaka.com/en/exhibitors/024/index.html.
33 Shinagawa *Higashi-Osaka Genki Kojo*, pp. 24–8.
34 Ibid.
35 Almost all of this section, including the diagram, is from www.m-osaka.com/en/exhibitors/024/products.html.
36 See www.rodan21.com/sangaku/connect.html.
37 Shinagawa *Higashi-Osaka Genki Kojo*, pp. 166.
38 Small and Medium Enterprise Agency *White Paper on Small and Medium Enterprises in Japan*, Tokyo: Japan Small Business Research Institute, 2005, pp. 261–2.
39 Ibid., p. 261.
40 Ibid., p. 262.
41 Ibid., p. 262.

7   University start-up ventures and clustering strategy in Japan

Akio Nishizawa

Introduction

According to the results of "The sun also rises: a special issue on Japan's economic revival," the latest survey conducted by *The Economist*, the prominent UK economic journal, "Japan is back. It is being reformed. It is reviving."[1] While we should admit this economic recovery has been based on the "accumulation of many small changes in politics, financial regulation, corporate law, public opinion, capital markets, corporate mores" as noted in the survey,[2] this new Japan has occurred in quite limited places, such as Tokyo, with a centralized effect of concentration of government offices and corporate headquarters, and the area of Nagoya, which has experienced the positive economic effects of the World-Expo having huge public spending for infrastructure and private spending of visitors, and Toyota, one of the world biggest car makers with high competitiveness. On the other hand, the survey also admitted that other areas outside of Tokyo "have suffered greatly, with shuttered-up streets and rising levels of poverty."[3] In fact, the report entitled *Regional Management under the Declining of Population*, published on December 2, 2005 by METI estimates that all cities other than Tokyo suffered severely from decreasing populations, and only 35 big cities out of 269 urban employment areas with a densely inhabited district of more than 10,000 people and a commuting rate of over 10 percent from their suburbs may be expected to increase their gross regional product (GRP), which is equivalent to GDP as forecast for the national economy in the year 2030 based on simulated projections.[4] New types of economic rejuvenation presented by the survey in *The Economist* cannot make whole regions in Japan grow to be prosperous with increasing employment as was the case before the 1990s.

The background to this restricted economic prosperity in Japan might be explained by the emergence of Asian countries, especially Mainland China, with which every Japanese region has competed severely, and huge fiscal deficits of central and local governments that cannot continue to support the local economy through public spending and agricultural subsidies against the pressure from Asian courtiers. Before the emergence of

Mainland China in the 1990s in the global marketplace, major Japanese manufacturing companies established their headquarters in Tokyo where they had close proximity to and better relations with the central government. They have also located their R&D centers with prototype manufacturing facilities in the suburbs of Tokyo. They have established their manufacturing facilities through a network of subcontracting SMEs in the local areas in Japan to exploit relatively low labor costs for manufacturing, where most of the workers can have jobs at manufacturing facilities of big businesses or SMEs and also as part-time rice farmers who can supplement their relatively low income at manufacturing facilities through subsidies from the central government for maintaining rice production. There has also been big public spending on construction activities that have created additional incomes for off-season farmers. All three major conduits for generating employment and income through paying back salaries and subsidies from big businesses and the central government to local areas have maintained bottom lines for economic prosperities in the local areas. Meanwhile, based on the rapid growth of big manufacturing companies in the post-Second World War era, the Japanese have enjoyed an increase in their incomes through the trickling down of profits through the above-mentioned conduits. After evaluation of the yen in the middle of the 1980s, this structure was forced to implement change. But in the late 1980s, the competitiveness with Asian countries was not so strong, fortunately, and the Japanese economy could absorb the bad effects of the high re-valuation of the yen without impeding the development of the bubble effect, which burst at the end of 1980s.

After the bubble burst, giving rise to huge bad debts in the financial sectors, and Mainland China emerged as a strong competitor, big Japanese manufacturing companies facing these problems tried to reduce their non-productive assets as much as possible through the restructuring and relocation of their manufacturing facilities from Japanese local areas to Asian countries, especially to Mainland China. According to the *Japanese SMEs White Paper 2005*, 51.6 percent of SMEs in Japan have been suffering from cut-price pressures from competing goods imported from outside of Japan. Out of this 51.6 percent, 70.6 percent of this competition is from Mainland China, 9.5 percent from ASEAN countries, 6.6 percent from South Korea, and 3.7 percent from Taiwan. And the number of SME companies decreased in the 1990s. If we take the index of company numbers at 100 in 1981, the index was 88.5 in 2001, and manufacturing sector companies have decreased to 70.3.[5] This could be a similar situation to that faced by the US in the late 1970s competing with Japan, when the US started new industrial and economic rejuvenating policies. Japan has been actively introducing similar policies, such as academia–industry technology transfer, university start-up ventures, and clustering in regions, learning from the experiences of the US.[6]

In this chapter, I will examine the feasibility of these policies by comparing the clustering strategies of two local areas: Austin, Texas, in the US and the city of Sendai, Japan.

The clustering economy in the US

The Cloning Silicon Valley policy

Faced with severe economic disease called "stagflation" in the late 1970s in the US, President Carter issued a "presidential message for innovation" in October 1979, which was the turning point from a long tradition of Keynesian demand management policy to supply-side policy for strengthening the industrial competitiveness to rejuvenate the US economy. Before issuing the presidential message, President Carter asked the Department of Commerce to establish a study team by inviting representatives from industry, labor, and academia to find a solution to save the US economy from stagflation. The study team had watched Silicon Valley, which had grown even in the stagflation era of the 1970s, and recommended the proliferation of Silicon Valley by introducing a "Cloning Silicon Valley policy," which would help regions to create high-tech industries by nurturing new technology-based ventures through exploiting advanced research results from prominent research universities and/or national research institutes in the regions. The Cloning Silicon Valley policy consisted of a new venture creation policy by the federal government and a new venture growth policy by the state and local government (see Figure 7.1).

There are two major risks for new venture creation. One is technology risk and the other is business risk. While start-up ventures have few management resources, they should realize break-through innovation for survival and higher growth. Usually start-up ventures have little R&D ability to create technological seeds for break-through innovations, but the real function of start-up ventures is to form a bridge between the seeds invented at research institutions and the market.[7] Therefore start-up ventures may be created, based on a fertile R&D environment requiring technological seeds from research institutions. In the case of Silicon Valley, Stanford University, Xerox PARC, and other governmental research institutions played an important role in supplying the seeds for start-up ventures. Based on these experiences in Silicon Valley, the US government introduced a new policy to open up research results from universities, private and government research institutions with funding from the federal government. This opening-up policy of federally funded research results was possible through new acts, the Bayh-Dole Act and the Stevenson-Wydler Act, passed by Congress in 1980.[8] But this policy changed drastically previous ownership of federally funded research results from government to private called the "Second Land-grant,"[9] which should have created new technology markets for the federally funded research results,

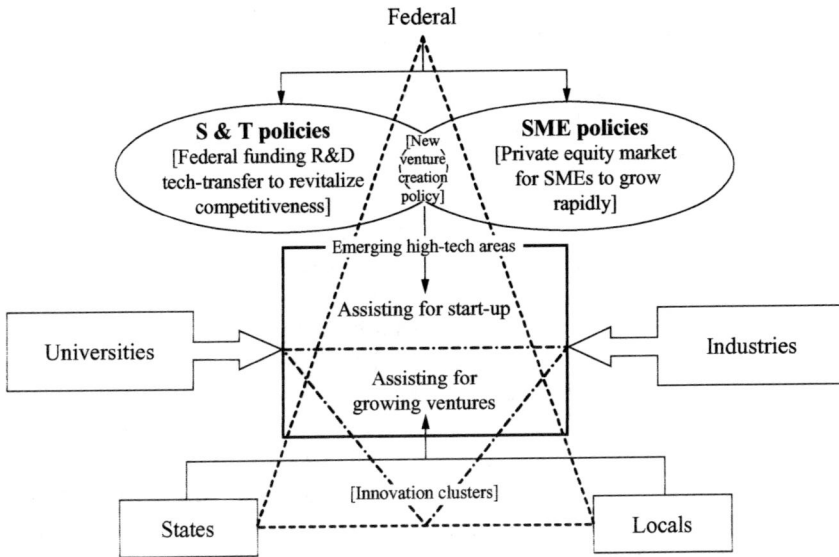

Figure 7.1 The Cloning Silicon Valley policy in the US
Source: The author.

but it took more time, well into the late 1980s, to be formalized under the academia–industry technology transfer policy.

The second obstacle for start-up ventures is business risk resulting from the lack of money and market. Start-up ventures find it quite difficult to obtain funding for their growth because they do not have the assets to use as collateral to offset the higher risks. There is also a big information gap between entrepreneurs, investors, and/or lenders. In Silicon Valley, there had emerged a specialized financial intermediary, VC companies, which played quite an important role in supplying risk money through equity investments, and in providing hands-on assistance to reduce the business risks. In the Cloning Silicon Valley policy, the federal government expanded VC investment activities by formalizing VC funds in the form of limited partnerships with changing the regulations of the Employee Retirement Income Security Act of 1974 (ERISA), which aimed to expand the flow to VC funds from pension funds. A limited partnership used as a VC fund is considered the most innovative funding scheme for venture financing.[10] Deregulation of ERISA has diffused the limited partnership fund in the US and has created a new financial market, the "private equity market."[11] In relation to expanding VC equity investment, the federal government deregulated the Initial Public Offering (IPO) and listing requirements at National Association of Securities Dealers Automated Quotations system (NASDAQ) market and decreased the capital gains tax rate drastically from 49 percent to 28 percent and finally to 20 percent in 1981.[12] All of these policies were aimed at facilitating risk money inflow to start-up ventures.

As for the lack of a market, new technology-based start-up ventures should supply the innovative goods to the market where they must face the difficulty of finding the "charter" customer[13] who can lead and certify the new innovative function. Based on the product life-cycle model for diffusion of innovation, the charter customer plays the most important role to introduce and diffuse the innovative goods successfully to the market.[14] In Silicon Valley, military demands played a similar function to the "charter" customer for start-up ventures introducing new technology-based break-through goods to the market.[15] In the Cloning Silicon Valley policy, SBIR, introduced by the Small Business Innovation Development Act of 1982, was expected to play the similar role that US military demand achieved for creating Silicon Valley in late 1960s.[16]

The federal government took the initiative to change the policy drastically as mentioned above by actively introducing the Cloning Silicon Valley policy, but its major activity in economic policy is to be limited to the macro-economic level under the US federal system. Micro-economic level policies must be entrusted to the state and/or local government. However, in the early days of the 1980s, the state and local government did not have any model to formally adopt such as the Cloning Silicon Valley policy. And, in reality, it seemed quite difficult for the state government to introduce the Cloning Silicon Valley policy, because it required the total change of previous policies for equal development of the state. Pursuing the Cloning Silicon Valley policy, the state government selected one site and focused on its development, which created political risks for the state governor. The state governor was required to take a more entrepreneurial role for risk taking in adopting the new policy to rejuvenate his or her state economy.[17] Yet, the expected results were vague if the governor dared to introduce a new policy without a successful model other than Silicon Valley, which seemed to be created spontaneously without any policies. All of these situations were totally changed by the success of the Microelectronics and Computer Technology Corporation (MCC) invitation, resulting in the "Austin Miracle" from the middle of the 1980s.

MCC and the "Austin Miracle"

In 1983, the City of Austin, Texas, was the winner of a site-selecting competition run by MCC out of fifty-seven cities in twenty-seven states, and successfully invited MCC headquarters to locate there. MCC was quite a unique research consortium established jointly by computer and semiconductor manufacturing corporations to recover competitiveness in the microelectronics and computer industry of the US against the perceived

Japanese threat, but it was bound to have authorization of its non-anti-trust intention from the Federal Trade Commission (FTC) because this type of R&D collaboration was against an anti-trust operation. The MCC insisted on its legitimacy by focusing on the pre-competitive R&D to compete against Japan-Inc., which strengthened its competitiveness through R&D collaborations in high-tech industries sponsored by the Japanese government, especially MITI. The MCC was the first private R&D consortium so it should have adopted a public operated style to survive the anti-trust decision by the FTC. The MCC invited Mr B. Inman, Deputy Director of the Central Intelligence Agency (CIA), to be the first President and began a nationwide site selection by showing site-selection criteria, including especially a properly run academia–industry–government collaboration in the regions. In the final stages of site selection, four cities, Raleigh-Durham, San Diego, Atlanta, and Austin, were left to present their merits at the final site-selection meeting. Austin's rank was fourth before the final presentation. Austin won to invite MCC, through its highly successful collaboration between university, industry, state, and local governments for achieving the site-selecting conditions. Austin being the winner gave great impact to both finalists and other cities with high-ranking research universities, which created a type of model for academia–industry–government collaboration for adopting a new policy to create another Silicon Valley in their regions by using research results from the university and research institutions. In 1988 Austin successfully invited SEMATECH (Semiconductor Manufacturing Technology initiative), an industry–governmental research consortium focusing on semiconductor manufacturing, and R&D centers of private companies based on the established academia–industry–governmental collaboration with highly talented research peoples and students.[18] All of these consortia, institutions, and human resources became the base for realizing the "Austin Miracle" in the 1990s (see Figure 7.2).

According to *R&D Collaboration on Trial*, academia–industry–governmental collaboration, the base for successful invitations of MCC, SEMATECH, and other R&D centers of private sectors, had been created intentionally by the prominent visionary, Dr G. Kozmetsky, who had great experience of being an entrepreneur, scholar, and initiator.[19] In the early 1980s, he advocated the "technopolis" concept to create a high-tech industry city with a high-quality of life based on regional research results through "collaborative individualism." He had organized individuals with power based in university, industry and government sectors to voluntarily encourage their organizations to provide social benefits that met with the goals of the MCC. His activities may be categorized as a "first-level influencer"[20] who had the status to make contacts with prominent persons among academia, industry, and government to further his idea, be acknowledged, and inspire cooperation to reach his highly respected vision. His organizing abilities among organizations and institutions created social

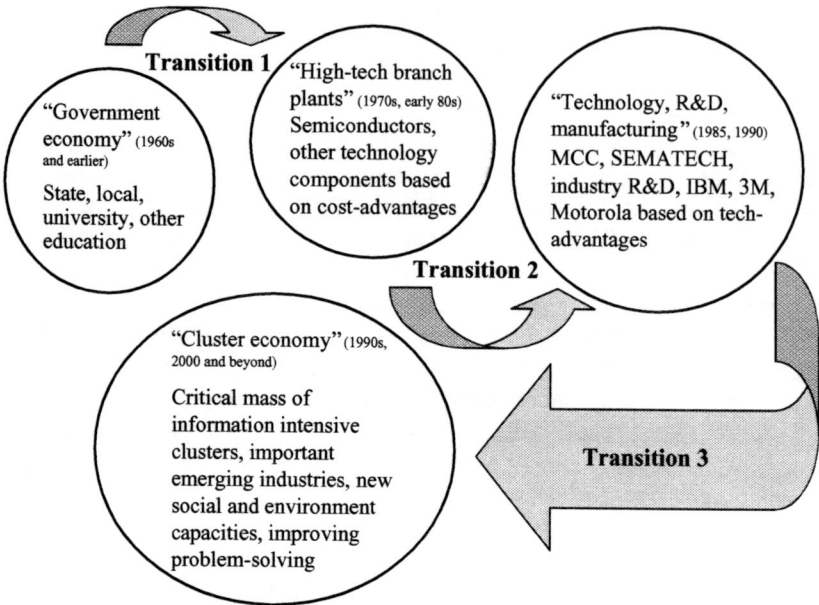

Figure 7.2 The "Austin Miracle"

Source: *Next Century Economy*, The Greater Austin Chamber of Commerce, USA, 1998, p. 4.

benefits that gave rise to information benefits and control benefits.[21] However, in the case of the establishment of MCC, SEMATECH, and R&D centers of private sectors, only information benefits could play an important role; they were not sufficient to create the "Austin Miracle."

The "Austin Miracle" had been realized by creating a new information technology industry through start-ups and accumulation of new technology-based ventures through exploiting the technological seeds emerging from the university, R&D consortia such as the MCC and SEMATECH, and the R&D centers of private sectors.[22] Start-up ventures even with high growth potentials are quite weak in the start-up phase where they need to be fully assisted to move from rapidly growing entities to be independent companies. These start-up ventures require the places for their business operations, development to commercialize the seeds of technology, risk money acquisition system through equity, highly talented human resources, and management consulting. The specialized place for meeting these requirements for start-up ventures is the incubator, which should be established and managed for effectively assisting ventures near the university and/or research institutions. The actual physical building for ventures to be housed is not enough for an incubator to thrive, but the incubator must have networks to provide the above-mentioned resources and services for start-up ventures. The incubator requires the incubating system. I categorize this incubating system as the "innovation cluster" (see Figure 7.3).

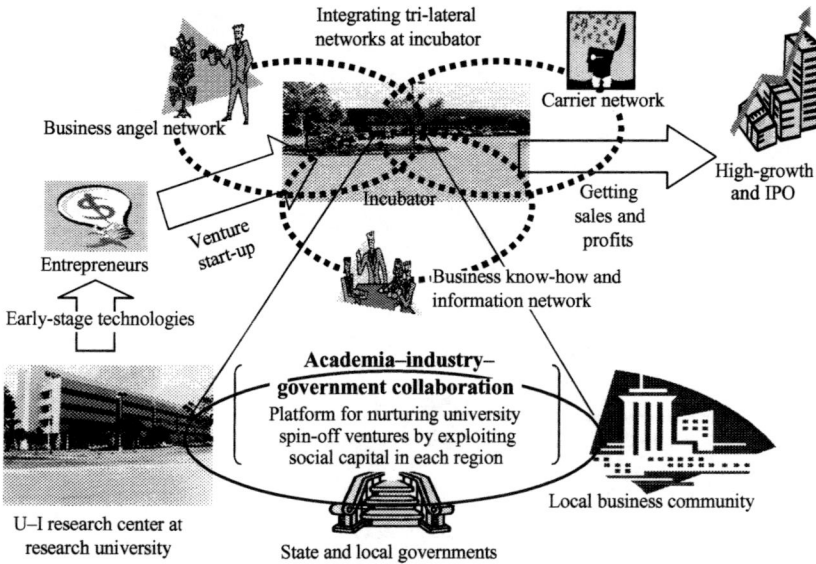

Figure 7.3 "Innovation cluster," key for the success of start-up ventures
Source: The author.

While Figure 7.3 is rather complicated, this model was created by our research on the case of Austin, Texas, for several years with Dr D. Gibson, from IC^2, the University of Texas, at Austin. At first, the university should establish a university–industry collaborative research center, such as MCC and SEMATECH, which was formalized under the National Cooperative Research Act of 1984 for technology incubating by introducing market needs brought by researchers from industry with generic technology invented in the university. When early-stage technology with high potential for commercialization may emerge from these collaborative research activities, which seems to be too early to be commercialized by the existing companies and tend to be impeded from continuing their further development, the researcher who invents this new technology with a good knowledge of the market potential will wish to market this early-stage technology through a new venture. To facilitate researchers to start their new ventures, incubators should invite the start-up venture after evaluating its feasibility and assist its growth while reducing business risk. The start-up venture can grow rapidly to be independent while being assisted in the incubator through networks of money, talented people, and management know-how within a number of years.[23] After graduating from the incubator, the start-up venture can grow rapidly to go public or to merge with existing companies through strategic alliances. The start-up venture can both increase the number of people employed and create income to the region (see Figure 7.4). All of these activities are also positively

Index

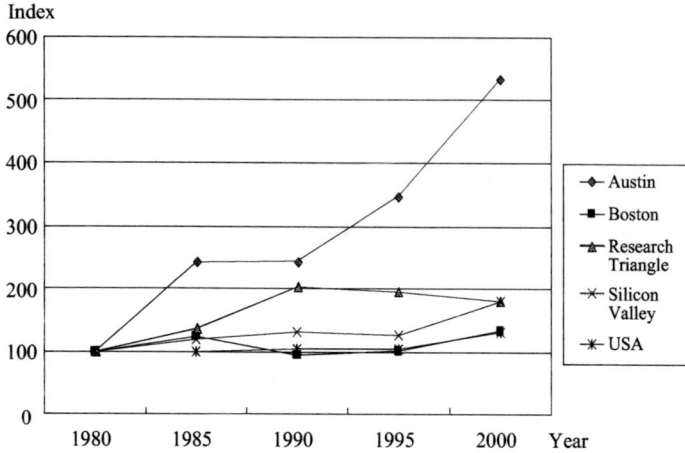

Figure 7.4 Incremental index of employment in selected US high-tech regions
Source: Austin Index issued by City of Austin, Texas, USA.

supported by the academia–industry–governmental network as the platform for new venture creation in the region.

The "Austin Miracle" has been brought about by the innovation cluster created by the first-level influencer, Dr G. Kozmetsky, assisted by working-level powerful staff members (called as second-level influencers) from academia–industry–government organizations. This should not be realized only by the networking of each related organizations with "information benefits," but also through leadership in creating the innovation cluster with "control benefits" to the region. The "Austin Miracle" has been highly regarded by other regions with high-ranking research universities in the US where similar strategies have been pursued based on the experiences of the "Austin Miracle." After spreading this strategy to other regions with some successful results, there are now many high-tech cities emerging in the US (see Figure 7.5). These newly emerging high-tech cities other than Silicon Valley and Boston Route 128 may have been the basis for the rejuvenation of the US economy in the 1990s.

Both European countries and Japan tried to pursue a similar direction of policy after poor economic conditions in the 1990s. The problem for European countries and Japan is whether they can create new economic growth based on the US-style clustering economy.

The "paradox of the global economy"

While information technology and other technological innovations have been said to lessen the economic borders of nations and allow them to become more global with ubiquitous knowledge in the "Flat World," the

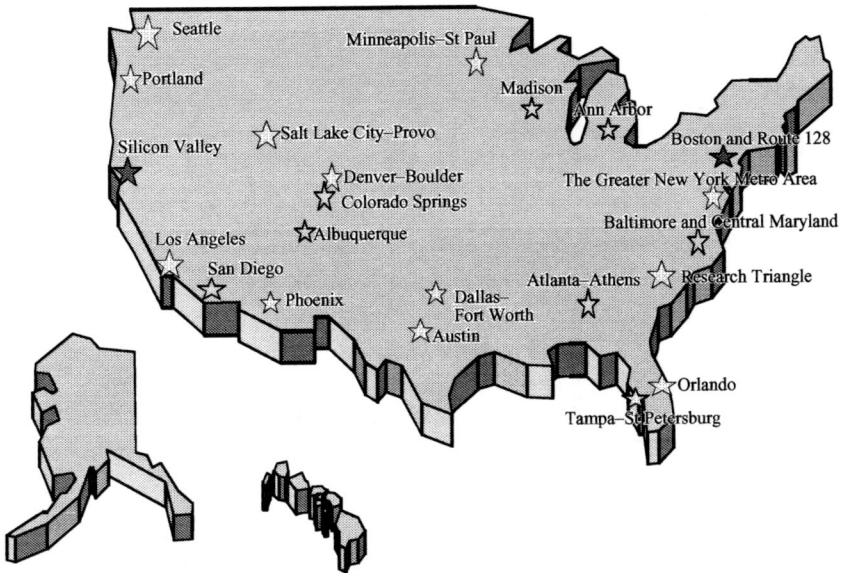

Figure 7.5 Major US high-tech cities

Source: E. Bolland and C. Hofer *Future Firms*, Oxford University Press, UK, 1998, p. 291.

hot spots creating innovations have narrowed to the city level as mentioned above in the case of the US. This situation is called the "paradox of the global economy" by Dr M. Porter.[24] It is the fact that there are increasing university start-up ventures and a "creative class" of highly talented people including entrepreneurs that was also growing in the 1990s after spreading high-tech cities based on the "Austin Miracle" (Figures 7.6 and 7.7). Information technology, bio-technology, medical and health care, new material, nanotechnology, and environmental technology are the bases for the growing high-tech cities, while research universities, government, and private research institutions are playing important supportive roles. This phenomenon might be explained by the importance of the innovation cluster, which can mostly rely on the tacit knowledge of highly talented people, or the "best and brightest." The innovation cluster can fill the gap between the start-up venture and resource suppliers. It can act as intermediary for risk money flow from angel investors through creating angel investing networks and VC to new ventures. It can play recruiting functions between ventures and highly talented people as the "weak tie" of job transfer network. It can gather such assisting professionals as patent attorneys, attorneys-at-law, certified public accountants (CPAs) and managements for ventures. All of these networks can work towards facilitating tacit knowledge on finance, recruiting, and assisting. And it is quite difficult for tacit knowledge to flow globally, even nationally, which

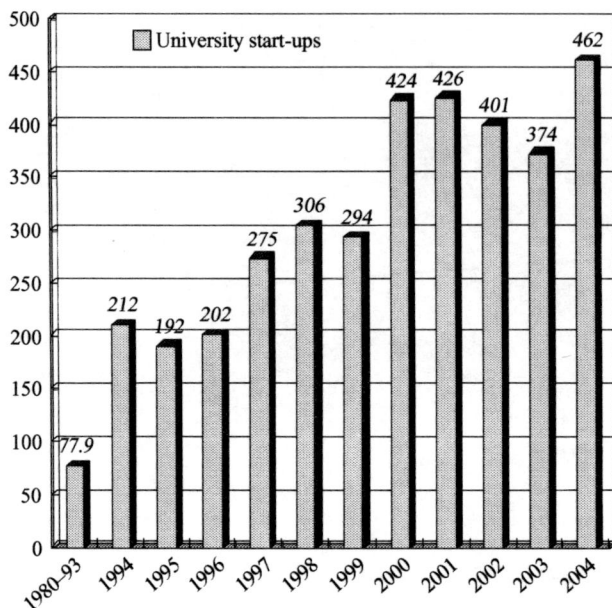

Figure 7.6 Growth of university start-up ventures in the US
Source: *AUTM Licensing Survey*, FY 2004, AUTM, USA, 2006, p. 28.

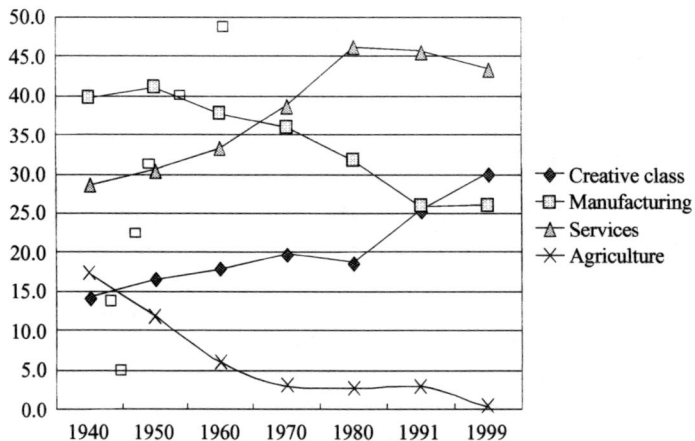

Figure 7.7 Changing employment in the US industrial sectors
Source: R. Florida *The Rise of Creative Class*, Basic Books, USA, 2002, p. 332.

can be transferred among the limited scope of the region through frequently gathering and moving of these people.

The importance of the changing of the technology, finance, and labor markets in the US for making the innovation cluster effective should also be pointed out. As already mentioned, the federal government has opened new technology markets by introducing new acts, such as the Bayh-Dole Act and the Stevenson-Wydler Act, for government-funded research results to be transferred from academia, government, and private research institutions to industry. As for the financial market, the federal government also opened new markets of private equity investment by formalizing limited partnership funds with the least regulated market for professional investors. This new private equity market has been expanded rapidly not only for venture financing but also for the restructuring of existing companies through leveraged buyout (LBO) and M&A financing (see Figure 7.8). In relation to these corporate restructuring activities, the "internal labor market," or "Fordism," has been drastically broken and even highly talented people with higher degrees cannot participate in big firms with higher income.[25] This drastically changing labor market forced the best and brightest to become entrepreneurs or join start-up ventures. While it is rather difficult to reduce the risk of firing by restructuring, risk in ventures can be controlled internally.

All of these changes in the technology, finance, and labor markets should prepare the clustering economy in the US, and these are the bases for the innovation cluster created around the university and research institutions in the regions. In the next step, academia, government, and industry in

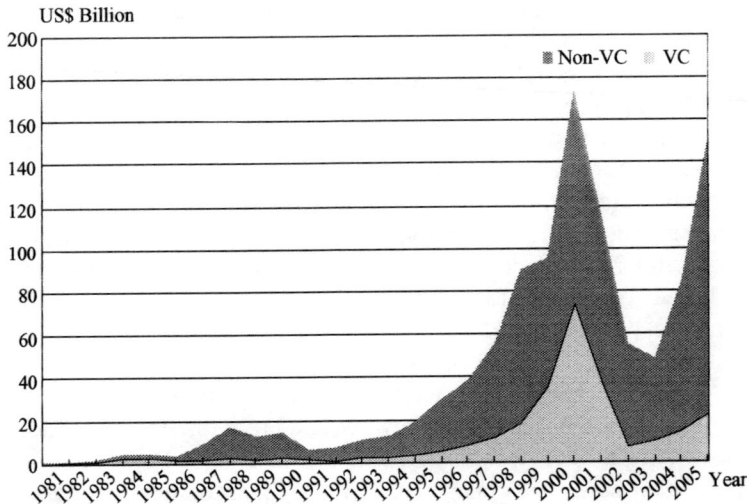

Figure 7.8 Fundraising for the private equity market

Source: Adapted by the author from *Dow Jones Private Equity Analyst*, various issues.

the regions should collaborate to facilitate new venture creations, acceleration of growth, and accumulation for high-tech industry through the innovation cluster. In this step, the visionary with high leadership abilities can take the initiative to make all the related organizations and institutions from academia, government, and industry participate in the innovation cluster as a first-level influencer. While the importance of the first-level influencer's initiative to create the innovation cluster by inviting related organizations and institutions in the region should be highly appreciated, this result might not be realized without any changes in the technology, finance, and labor markets brought about by the federal government. The Cloning Silicon Valley policy by the federal government coupled with the creation of the innovation cluster in the regions played an important role to save the US economy from the stagflation in the 1970s and to rejuvenate it with the cluster economy in the 1990s.

Now we can summarize the evolution of the US economy based on the unique model of the Triple Helix introduced by Dr H. Etzkowitz. The Triple Helix model can explain the change in relationships among academia, government, and industry explicitly step by step (see Figure 7.9). While this model is important in explaining what process should be necessary for changing the system and how the relationship between academia, government, and industry has evolutionally changed from Etatism to the Triple Helix, it cannot explain clearly what has been changed between each step, which should be complemented by the factors found by our research.

The Etatistic model represents the economic model operated during the Cold War, where the government strictly controlled the markets and its participants to sustain economical growth and prosperity. But this model could not work well because of oil crises and strong pressure of global competitiveness. Stagflation appeared in the US and was the signal for not keeping this model. At the same time, the Etatistic model was applied to twentieth-century-type industry that involved intensive capital and mass labor manufacturing, but this form of industry also needs to be changed to twenty-first-century high-tech industry that requires knowledge-intensive investment. Then the Etatistic model should be changed drastically to create the new model for creating twenty-first-century-type of knowledge-based economy. But the Triple Helix model tells of the important implication that, before creating the new model (the Triple Helix model), the Etatistic model can be broken up by introducing Laissez-faire model because the market can change drastically based only on economic rationality for preparing the new model. From the Etatistic model to the Laissez-faire model, technology, finance, and labor have been the central fields that should be changed first through introducing a market system as mentioned above. These three fields were the core of the Etatistic model controlled by the government until the early 1980s. As previously mentioned, the US government opened up new markets for federally

Etatistic model   **Laissez-faire model**   **Triple Helix model**

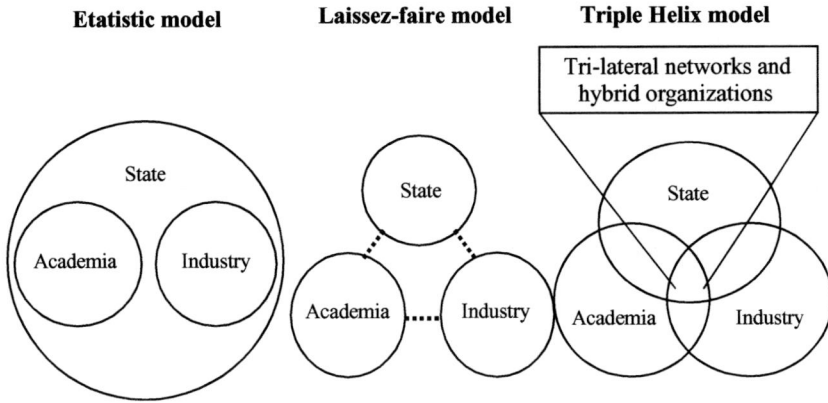

Figure 7.9 From Etatistic model to Triple Helix model

Source: H. Etzkowitz and L Leydesdorff "The dynamics of innovation: from national systems and 'Mode 2' to a Triple Helix of university–industry–government relations," *Research Policy*, No. 29, Elsevier, p. 111.

funded research results, risk money supply, and high-talented people in 1980s. However, these do not trade well in the markets because there are too big information gaps between suppliers and demands to be filled in the normal market mechanisms. The Laissez-faire model can break the Etatistic model for extracting the core factors, but it cannot create the new model with effective mechanism to solve the information asymmetries for trading the above-mentioned core factors in the markets.

The Laissez-faire model should be changed to the Triple Helix model where these information asymmetries can be reduced through "tri-lateral networks and hybrid organizations." But according to the Triple Helix model, the core structure and function of the "tri-lateral networks and hybrid organizations" cannot be fully explained. The core structure and function of the Triple Helix model can be clearly understood by using our Innovation Cluster concept. The core of Triple Helix model, "tri-lateral networks and hybrid organizations," can be categorized as the Innovation Cluster. And the problem of the "paradox of the global economy," can also be explained based on the hypothesis where the Triple Helix model is the Innovation Cluster. As mentioned above, tacit knowledge, which cannot be coded and separated from the people involved, plays quite an important role in the innovation cluster for facilitating start-up ventures based on the advanced research results from collaborative research activities in the university or research institute. Therefore, the people with high tacit knowledge should move on to start their own ventures or join those ventures in the regions where their tacit knowledge can be utilized gainfully by these ventures accumulated in the similar technological sector. The scope for moving people with high tacit knowledge[26] can make some geographical limitation around the university or research institute that provides the technological seeds and talented people in the similar fields. Furthermore, risk money providers/business angels and VC can assist new ventures located in the vicinity. This is the background of creating the "paradox of the global economy," which was the key for success to rejuvenate the US economy in the 1990s and beyond.

Clustering strategy in Japan

From "technopolis" to clustering policy

The Cloning Silicon Valley policy to rejuvenate the US economy, which started in the early 1980s, was composed of a new venture creation policy by the federal government and a new venture growth policy by the state and local governments. These two policies have been integrated effectively, after the city of Austin, Texas, won the MCC site selection in 1983, where there was a unique and quite effective collaboration between university, industry, state, and local government under the leadership of a visionary as the first-level influencer, who proposed the creation of "technopolis" there. From this surprising victory of Austin within the context of a possible losing game, it became clear for regions with high-ranking research universities and research institutions that to organize such effective academia–industry–government collaboration for saving the regions from stagflation, this had to be done through creating high-tech industries with the accumulation of new ventures based on the research results from the university or research institutes located in the regions. These activities have proliferated to both losing finalists for the MCC site selection and other regions in the US. These activities have created such prominent results in the regions as to rejuvenate the US economy in the 1990s, as mentioned above.

In opposition to these new developing policies in the US, Japanese "technopolis" policy, promulgated in the early 1980s, could not have had a similar impact on the Japanese economy, especially the regional economy in Japan. At first, the Japanese government, especially MITI, planned to establish one or two "technopolises" as the symbol for changing the Japanese economy from mass manufacturing to a knowledge-based economy after surviving two oil shocks, which marked the end of mass manufacturing based on cheap oil. But, when the Japanese government declared the "technopolis" policy, many governors of prefectures, the meso-regional government in Japan, rushed to lobby MITI to invite "technopolises" to their prefectures. Lobbying from prefecture governors was so strongly assisted by politicians that MITI could not deny them, accepting twenty-six proposals to create "technopolises" in these prefectures. In those days there were severe concerns about the growth potential of local economies due to the oil shocks. Then the Japanese "technopolis" policy was focused on filling the gap of infrastructures for creating innovations between the metropolitan areas and local areas.[27]

Officials from MITI tried to introduce the Silicon Valley phenomenon to Japan, however, they decided that it was impossible for Silicon Valley-generated spontaneous individualism to be transferred to Japan. Then they thought it necessary to have initiatives to build the facilities for academia–industry collaborative research, incubators, science parks and related hardware, which were thought to be a kind of incentive for inviting the manufacturing facilities of big firms, especially from both electronics and precision manufacturing. MITI and other governmental officials visiting Silicon Valley could only grasp the superficial phenomena such as start-up ventures, incubators, science parks, Stanford University, and VC. They did not understand the true structural changes pursued by the US to survive the stagflation, which focused not only on the facilities and hardware but also on legal issues, networking, and particular ways of conducting business. In the Japanese "technopolis" policy, the central government had poured large amounts of money into constructing many high-tech-like buildings in the assigned places such as the "technopolis," as a new type of public spending. In the end, Japanese "technopolis" policy could not strengthen R&D abilities in the regions, but only invited some manu-facturing plants of big businesses from the fields of electronics and precision sectors to construct once again through public spending many glass-worn unique style buildings.

Unfortunately Japanese "technopolis" policy could not fill the power gap for creating innovations between the metropolitan areas and the regions. For example, if the region had the power of creating innovations, this should have been accompanied by the distribution of patent attorneys, but they are still concentrated in the metropolitan areas (see Figure 7.10). While the result of the Japanese "technopolis" policy was clearly limited toward the last phase of the policy in the late 1980s, most people did not seriously care about it because the Japanese economy descended into the bubble in those years. Most people were caught up in the euphoria and thought there would be a continuous growth of the economy. "Japan as No. 1" was the symbolic expression showing the mood of the Japanese in the late 1980s. However, the Japanese were forced to face the severe situation of a long recession, called the "Lost Decade" after the bubble burst in the 1990s, as the result of doing nothing during the bubble econ-omy. The regions especially were the most damaged economically as mentioned above.

Then the Japanese government, especially MITI, started conducting research again as to how the US economy had been rejuvenated in the early 1990s, regaining its competitiveness. They found the academia–industry–government collaboration backed by the Bayh-Dole Act and clus-tering high-tech industries such as information technology, bio-technology, new materials, nanotechnology, new energy and the environment.

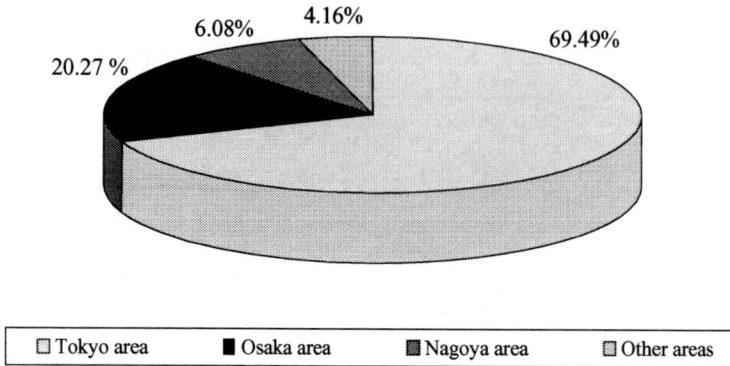

Figure 7.10 Regional distribution of Japanese patent attorneys (May 31, 2006)
Source: Japanese Association of Patent Attorneys.

University–industry collaboration

MITI started actively introducing a new policy of academia–industry collaboration to create high-tech industries in the regions against the hollowing out of their industrial bases. The university–industry collaboration, technology licensing from university to industry, and the clustering economy are the key phrases that have become quite popular from the end of the 1990s to save the Japanese economy from the "Lost Decade." There have been a lot of policies to facilitate academia–industry collaboration in Japan. In 1998, the Japanese government, especially initiated by MITI (now METI) and the Ministry of Education (now MEXT),[28] passed a new law, "Promoting University-Industry Technology Transfer" (the TLO Law), to activate the formal technology transfer through universities' related TLOs with financial subsidizing. After passage of the TLO Law, the complementary new laws were enacted successively: "Law on Special Measures for Revitalizing Industrial Activities" in 1999 (the 1999 Law), and "Law for Enhancing Industrial Technology" in 2000 (the 2000 Law). Then in 2001, Takeo Hiranuma, the Minister in charge of METI, launched the "New Plan for Creating 1,000 University Start-up Ventures within 3 years," in which university start-up ventures were expected to play a key role in rejuvenating the Japanese economy through diversifying its industrial structure from mass manufacturing to high-tech based on the advanced research results in new materials, information, and bio-technologies performed at the universities.[29]

Then MEXT started actively expanding university–industry collaboration because this policy became so popular in the central government and could be allocated a rather bigger budget. Just before the structural change of the national universities, MEXT asked both the national and the private universities to propose their own patent management organizations, which

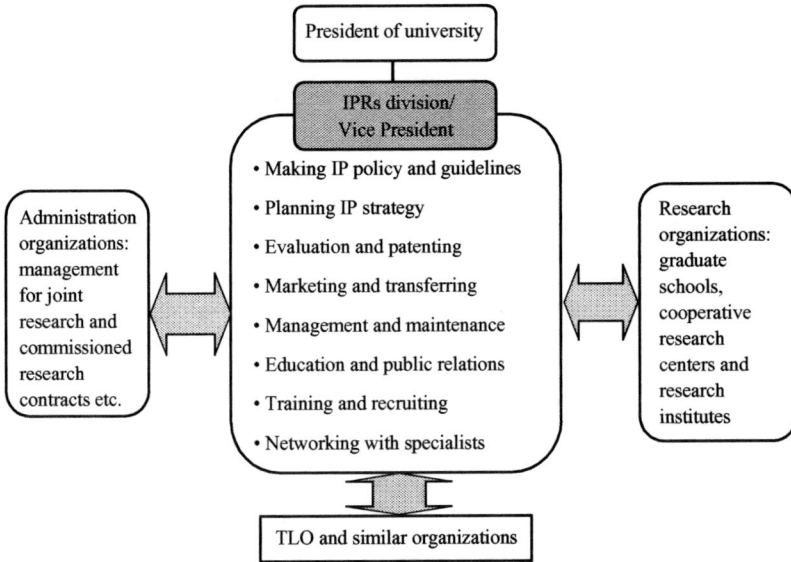

Figure 7.11 Basic idea for new IPRs divisions in universities as proposed by MEXT

Source: Translated by author from *Monbu Kagaku Kyoiku Tushin*, No. 69, February 2003, Tokyo, Japan, p. 25.

were selected to be awarded for realized the selected proposals by MEXT (see Figure 7.11). While there were already TLOs affiliated to the major universities, they tried to obtain awards by proposing IP divisions. There are now thirty-four universities awarded (see Figure 7.12). In reality, the universities having both TLO and IP divisions had a difficult time in integrating these two organizations for smoother technology licensing activities after establishing the IP division. This dual system seems to be ineffective for transferring the technology from the university to industry. And as for the "1,000 University Start-up Ventures" policy, the target was cleared at the end of FY 2004 with the number of 1,112 (see Figures 7.13, 7.14, and 7.15). These results show very clearly how the policies have been realized superficially, but there are lots of problems existing, as mentioned later.

Before explaining the problems to be solved, I want to show how the region and the university have reacted to these policies through one of the cases in the city of Sendai and Tohoku University in the following section.

Case of region: Sendai City and Tohoku University

The city of Sendai is the main city in the Tohoku region, which is composed of six prefectures in the northern part of Honshu. The population

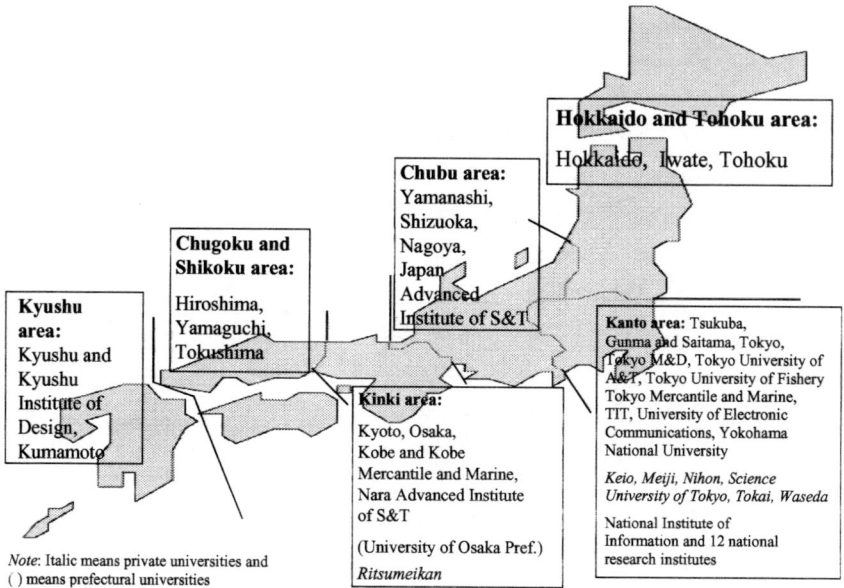

Figure 7.12 Thirty-four university IPRs divisions awarded by MEXT

Note: Italic is used to indicate private universities and parentheses to indicate prefectural universities.

Source: MEXT press release, July 15, 2003, translated by author.

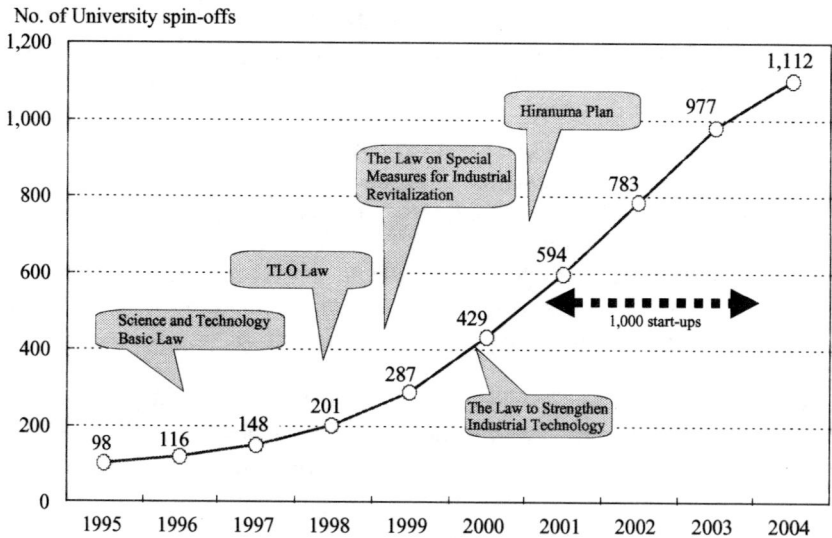

Figure 7.13 Growing numbers of university start-up ventures in Japan

Source: *Survey on University Start-up Ventures FY 2003*, METI, 2005.

For the past few years, university spin-offs in the biotechnology/medical fields have been increased. It is high ratio of the establishment of university spin-offs in the areas where universities hold strong research abilities that seems to indicate the roles of spin-off ventures as the alternative ways to market early-stage university research results in Japan.

Industrial sectors where university spin-offs established in each fiscal year

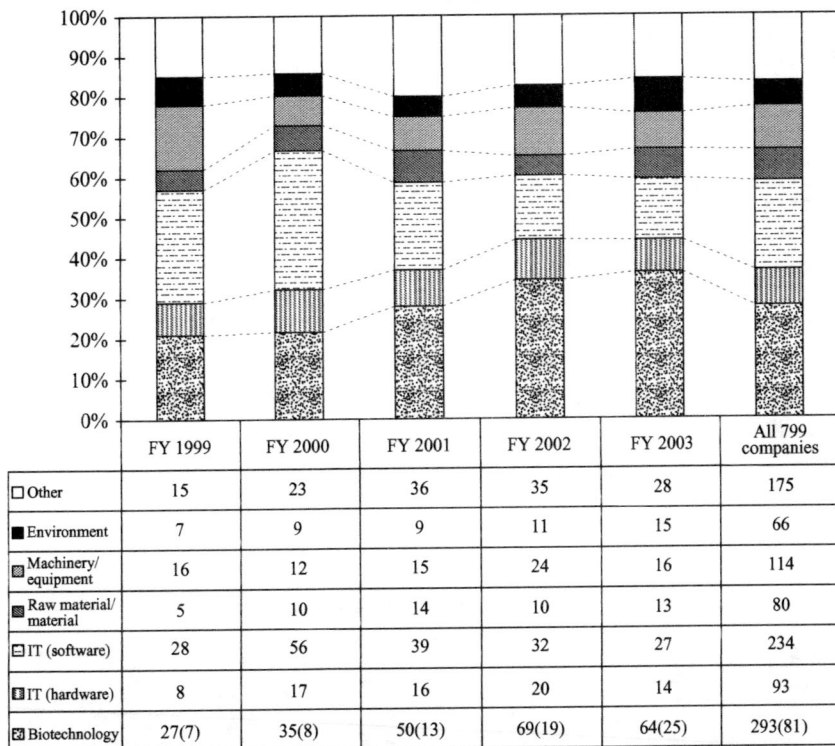

	FY 1999	FY 2000	FY 2001	FY 2002	FY 2003	All 799 companies
☐ Other	15	23	36	35	28	175
■ Environment	7	9	9	11	15	66
Machinery/ equipment	16	12	15	24	16	114
Raw material/ material	5	10	14	10	13	80
☐ IT (software)	28	56	39	32	27	234
☐ IT (hardware)	8	17	16	20	14	93
☒ Biotechnology	27(7)	35(8)	50(13)	69(19)	64(25)	293(81)

Figure 7.14 Where university start-up ventures have been born

Source: *Survey on University Start-up Ventures FY 2003*, METI, 2005.

of Sendai is just one million. It is a typical "branch economy" city, which means "almost big" and medium-sized firms have their branch offices there as the regional center in the Tohoku region. However, after the bubble burst, these branch offices have reduced their sizes and personnel, mainly due to severe cost reduction pressure to survive the long recession in the 1990s, and this drastic scaling down of branch offices is possible because of highways and bullet train networks being completed, which are the results of huge public spending starting in the early 1970s. The number of employment and offices in Sendai has been static during the late 1990s and decreased a little bit from the early 2000s. The economic conditions of Miyagi Prefecture, where the city of Sendai is located, have been also quite severe from the middle of the 1990s. Some manufacturing

Although leading universities such as ex-imperial universities, and Waseda and Keio lead others in cumulative numbers, regional universities have made good results in FY 2003, making it evident that university spin-offs have been spreading steadily.

The top ten universities related to university-based start-ups (cumulative)

Ranking	University name	Number of companies	Number of start-ups established in FY 2003
1	Waseda University	50	3
2	University of Tokyo	46	1
3	Osaka University	45	4
4	Kyoto University	40	7
5	Tohoku University	35	2
6	Keio University	31	2
7	Hokkaido University	26	7
8	Kyushu Institute of Technology	25	4
9	Kyushu University	23	0
10	Tokyo Institute of Technology	22	1

The top ten universities related to university-based start-ups (single year)

Ranking	University name	Number of companies	Number of start-ups established in FY 2003
1	Hokkaido University	26	7
1	Kyoto University	40	7
3	Tokyo University of Agri-culture and Technology	18	4
3	Nagoya University	19	4
3	Osaka University	45	4
3	Yamaguchi University	16	4
3	Kagawa University	8	4
3	Ehime University	6	4
3	Kyushu Institute of Technology	25	4
10	Iwate University	5	3
10	Kobe University	15	3
10	Kumamoto University	6	3
10	Waseda University	50	3

Figure 7.15 Number of university start-up ventures in FY 2003

Source: *Survey on University Start-up Ventures FY 2003*, METI, 2005.

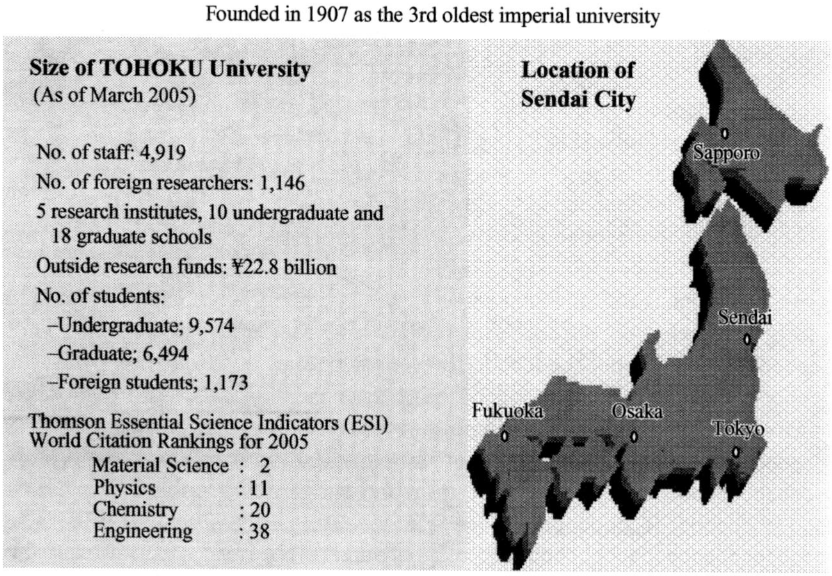

Founded in 1907 as the 3rd oldest imperial university

Size of TOHOKU University
(As of March 2005)

No. of staff: 4,919
No. of foreign researchers: 1,146
5 research institutes, 10 undergraduate and 18 graduate schools
Outside research funds: ¥22.8 billion
No. of students:
–Undergraduate; 9,574
–Graduate; 6,494
–Foreign students; 1,173

Thomson Essential Science Indicators (ESI) World Citation Rankings for 2005
Material Science : 2
Physics : 11
Chemistry : 20
Engineering : 38

Location of Sendai City

Figure 7.16 Tohoku University in Sendai City

Source: Tohoku University.

plants of big corporations have relocated from there to Asian countries, mainly to Mainland China.

Just before starting the new academia–industry collaboration policy, Tohoku University tried to establish a new facility for collaborative research with industry. Tohoku University is the third oldest national university, and its engineering school and research institutes on ferro-material and telecommunication have created a high level of research results (see Figure 7.16). The newly established university–industry collaborative research center is called the New Industry Creation Hatchery Center (NICHe). NICHe has two divisions, the Liaison Office and Industry Creation Section (ICS). Collaborative research with industry for technology application is done in the ICS. In ICS, there were thirteen projects in the national strategic technology fields (see Figure 7.17). After promulgating the TLO Law, Tohoku University established its TLO. Based on the US experience, there are three types of relations between the TLO and the university (see Figure 7.18), of which the Boundary model is estimated to work better than other two extreme models. The Internal model is usually governed by the university's administrators and it is difficult to treat the person in charge of tech-transfer as specialist. It put the priority not on tech-transfer, but getting outside research funding. The External model, on the other hand, put priority on profit, and though the productivity of tech-transfer may be good, its policy may not be in line with that of the university, and neutrality and the public nature of the university may also be sacrificed. On that day, when the TLO was established, the national universities in Japan did not have any legal status, which cannot have any direct relation with the TLO. It was concerned that the TLO was established by the third party to the university without having any control by the university. Then TLO was planned to be established as a joint stock corporation with faculties' share holding, which can easily undertake the business of technology transfer between the university and industry in line with the university's policy. Tohoku Techno Arch Co. Ltd (TTA) is the name of the TLO of Tohoku University. TTA was an approved TLO along with another three TLOs, which could be subsidized from METI for the first five years capped at 30 million yen per annum.

Before April 2004 when the national universities were changed with their whole structures to become national university corporations, TTA bought the patent filing right from individual faculty members or researchers and began to find suitable candidates to be transferred to become patented inventions. After collecting incomes and royalties, the TTA distributes a quarter of all earnings generated from the invention (net of cost) to each of the following four parties: the inventor, his or her laboratory, the university, and TTA (see Figure 7.19). While the technology licensing managers of the TTA tried to actively locate the candidates in the Sendai area, it was quite difficult because of the scarce number of manufacturing companies requiring advanced technologies resulting from

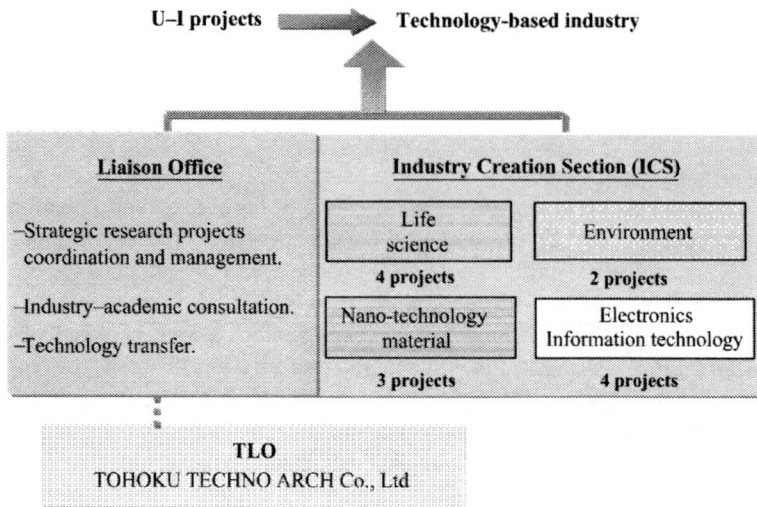

Figure 7.17 NICHe of Tohoku University
Source: NICHe.

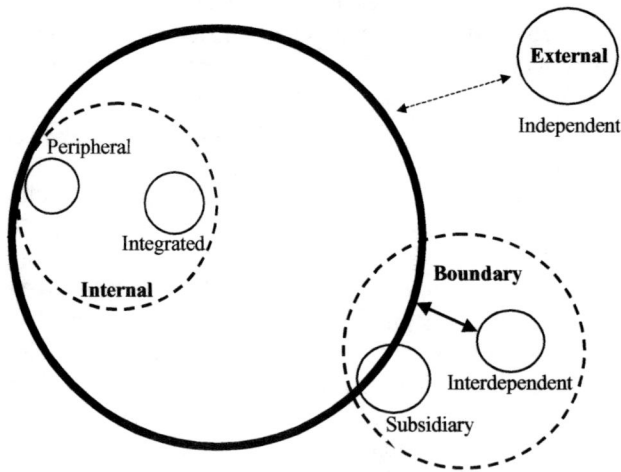

Figure 7.18 Models for organizing TLO in the US

Source: G. Makin "Organizing university economic development: lessons from continuing education and technology transfer," in J. P. Pappas (ed.) *The University's Role in Economic Development: From Research to Outreach*, Jossey-Bass Publishers, 1997, p. 33.

Research result — Licensing

University researcher

TLO
Tohoku Techno Arch Co., Ltd

Private companies

Distribution, feedback — Income

Patent application — Right maintenance

Establishment Nov. 1998
Approved Dec. 1998
Shareholders 254 from
    researchers at Tohoku area
Paid-in Capital ¥94,450,000

Patent office

Distribution rule

Inventor	1/
Laboratory	1/
University	1/

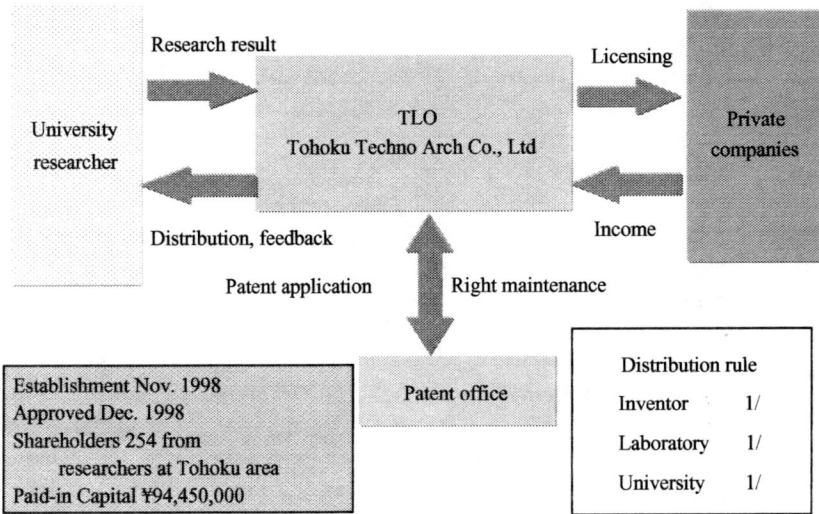

Figure 7.19 Technology transfer at Tohoku University
Source: Tohoku University.

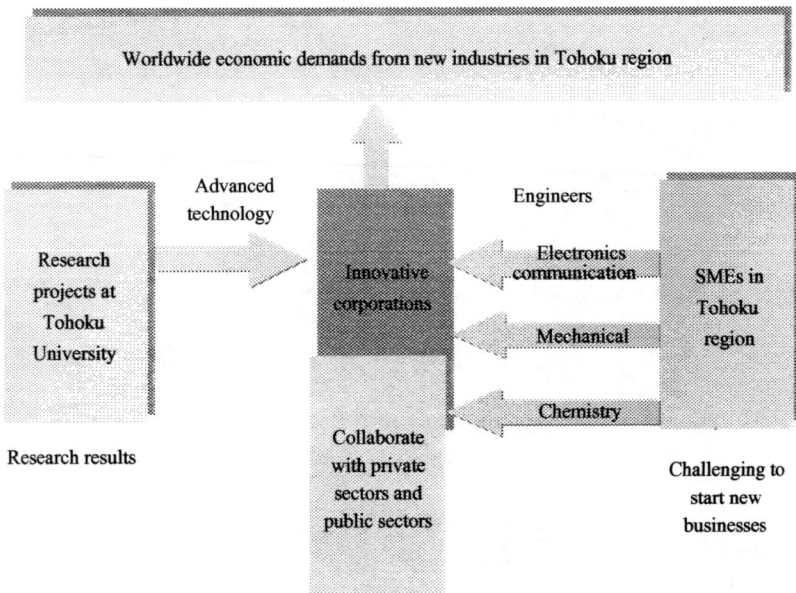

Worldwide economic demands from new industries in Tohoku region

Advanced technology — Engineers

Research projects at Tohoku University

Innovative corporations

Electronics communication

SMEs in Tohoku region

Mechanical

Research results

Collaborate with private sectors and public sectors

Chemistry

Challenging to start new businesses

Figure 7.20 The Sendai Model
Source: NICHe.

the university. Then NICHe tried to establish a new company to exploit the technology that is carved out of the strong expertise from existing SMEs in the region with the assistance from the TTA. We call it the "Sendai Model" (see Figure 7.20). Tohoku University also tries to organize the quasi-innovation cluster surrounding the university.

In the "technopolis" policy in the early 1980s, as mentioned earlier in this chapter, Sendai was one of twenty-six selected "technopolis" sites,[30] but it was strange there were no real relations with Tohoku University. The incubator and collaborative research facility was located far from the university.[31] Therefore it was necessary for the university to establish all the required facilities by providing risk money to assist in developing expertise around the university. The university has asked central and local government and economic organizations to cooperate in creating assisting networks for university start-ups. Figure 7.21 shows the current target of assisting networks in a new developing campus at Aoba hill, where new research park will be opened. Tohoku University ranked fifth in FY 2004 in starting ventures from the university, but there are no IPOs or trade sales as there is still not enough power to create high-tech industry in the Sendai area. The university should continue to work hard in assisting and accumulating university start-up ventures to create some good results while facing a lot of difficulties for providing of risk money and attracting highly talented people in Sendai.

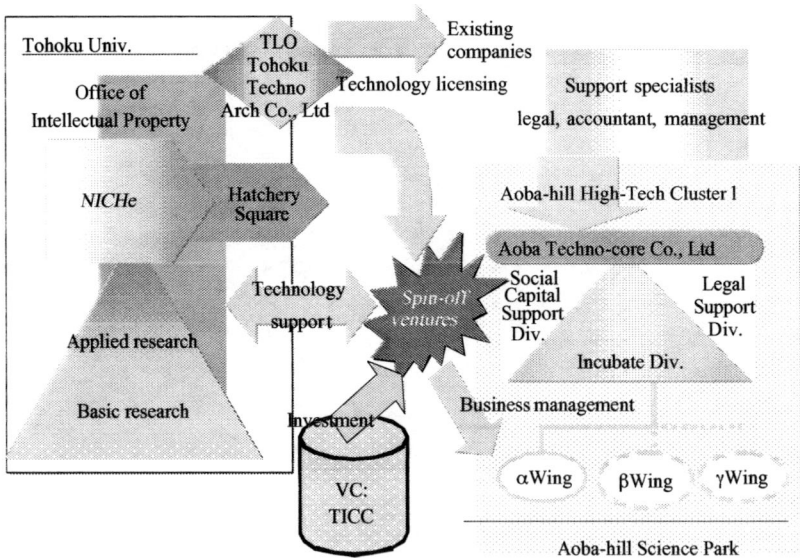

Figure 7.21 Aoba hill high-tech project
Source: NICHe.

Conclusion: new challenges for the Japanese economy

While the Japanese government, especially METI and MEXT, has actively pursued the US-type clustering economy by accumulating new technology-based ventures through the new collaboration of academia–industry–government in the regions where the universities are located, the biggest gap appeared in the industrial production index among regions. As for the industrial production index in October 2005, while Hokkaido is only 91.6, the Tokai area, vicinity of Nagoya, shows 121.9 comparing the base year of 2000 (2000 = 100).[32] The gap becomes more than 30, historically high in Japan. The ratio of jobs is also showing a big gap between Tokai and Hokkaido. Why is the economic performance in Tokai so good? The answer is Toyota, one of biggest car makers and with the highest profitability. This means that the Japanese economy still heavily relies upon big businesses, and the areas with headquarters of big businesses can enjoy economic prosperity.

In contradistinction to these good performances of restructured big business, the performance of university start-up ventures in Japan is not good enough to boost the regional economy by bringing new high-tech industry, comparing them even with the normal start-up SMEs (see Figure 7.22). The lack of risk money and highly talented people (the best and brightest) are the most important reasons why university start-up ventures do not grow faster than their counterparts in the US (see Figure 7.23). While the Japanese government has genuinely changed technology-transfer systems between the university and industry, and also university structure, it has still taken the policy to maintain the existing system in financial and labor markets. The Japanese financial policy has only been focused on saving the banks from bad assets problems during the 1990s without any measures for facilitating risk money to new markets such as the "private equity market" in the US. The result of this policy shows the lack of growing VC industry. Japanese VC investment activities are still very tiny compared with the US and Europe (see Figure 7.24); and 53.2 percent of individual financial assets remains as deposits, while only 6.6 percent is in equity holdings in the third quarter of 2005. The total amount of individual financial assets at that time was 1.43 quadrillion yen.[33] In the capital access index and country rankings of the *Capital Access Index of 2005* published by the Milken Institute, Japan is nineteenth, behind Malaysia, Spain, and Chile.[34]

Therefore there is a mixture of policies in Japan. One is to pursue the US-style Cloning Silicon Valley policy, and the other is to maintain the status quo. This mixture still exists in the minds of people. It seems the people living in metropolitan areas actively supported Koizumi's structural changes for pursuing the US-style economic rejuvenating policy while there was strong opposition in the regions according to the latest general election results. This is quite an interesting contradiction in the

128   *Akio Nishizawa*

While university-based start-ups have been increasing in number on a nationwide scale, their sales are far below the average sales of small and medium-sized companies. The impending challenges to be addressed are to stimulate the growth of university-based venture companies and substantially connect the activity to the economic value of Japan.

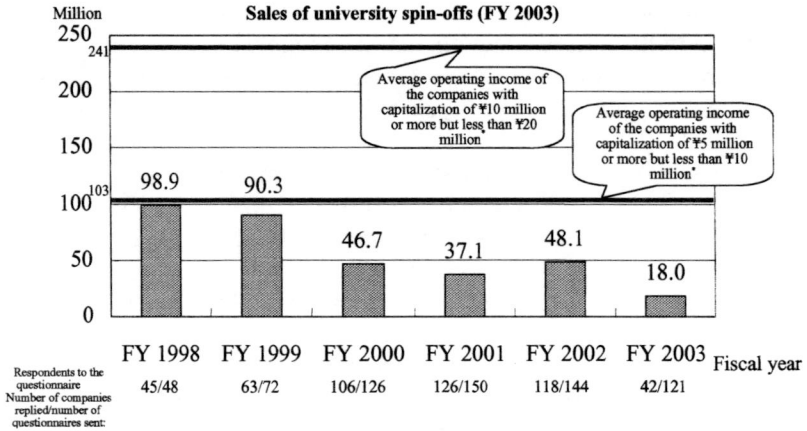

Sales of university spin-offs (FY 2003)

Million

Average operating income of the companies with capitalization of ¥10 million or more but less than ¥20 million

Average operating income of the companies with capitalization of ¥5 million or more but less than ¥10 million*

	FY 1998	FY 1999	FY 2000	FY 2001	FY 2002	FY 2003
Sales	98.9	90.3	46.7	37.1	48.1	18.0

Respondents to the questionnaire
Number of companies replied/number of questionnaires sent:

| 45/48 | 63/72 | 106/126 | 126/150 | 118/144 | 42/121 |

Figure 7.22 Lack of growth potential for university spin-offs in Japan
Note: *Investigated by the National Tax Administration Agency (FY 2002).
Source: *Survey on University Start-up Ventures FY 2003*, METI, 2005.

University-based start-ups have been suffering especially from the scarcity of human resources necessary for R&D, sales and marketing that need to be conducted additionally. In fact, two-thirds of the university-based start-ups in operation feel the shortage of talented employees.

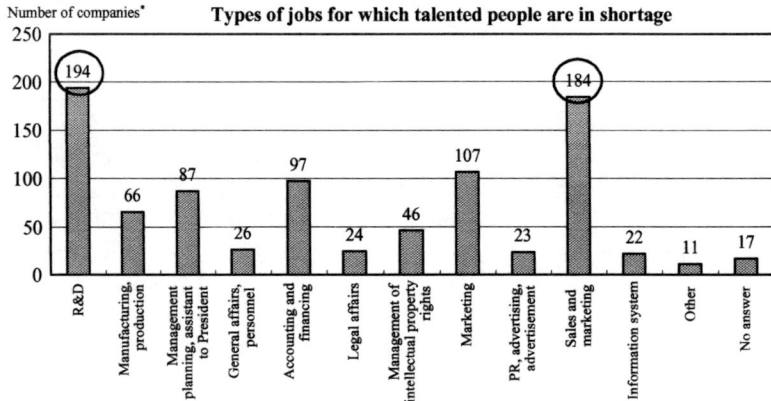

Types of jobs for which talented people are in shortage

Number of companies*

R&D	Manufacturing, production	Management planning, assistant to President	General affairs, personnel	Accounting and financing	Legal affairs	Management of intellectual property rights	Marketing	PR, advertising, advertisement	Sales and marketing	Information system	Other	No answer
194	66	87	26	97	24	46	107	23	184	22	11	17

Figure 7.23 Lack of "brightest and best" for university spin-offs in Japan
Note: *Of the 799 university-based start-ups, 362 respondent companies constitute the parameter.
Source: *Survey on University Start-up Ventures FY 2003*, METI, 2005.

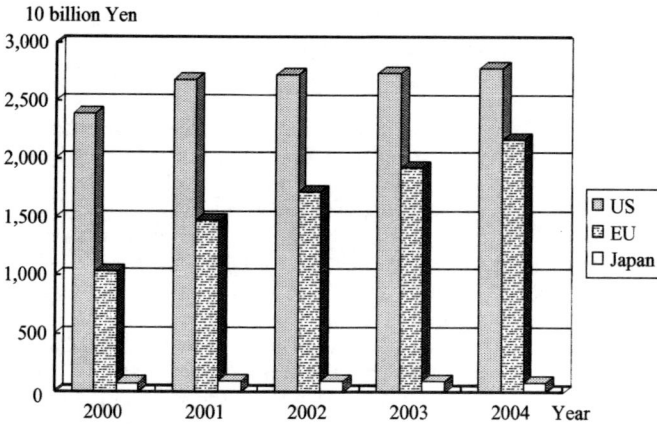

Figure 7.24 VC investment balances in the US, EU, and Japan
Note: Exchange rate, $1 = ¥107, Euro1 = ¥139.
Source: Venture Enterprise Center, *VEC VC Survey Report 2005*, VEC, Japan, 2005.

regions, but the concerns of people in the regions about the drastic reduction of pouring money from the central government in the forms of public construction and rice price supporting need to be acknowledged. Yet, it also needs to be addressed the fact that these policies maintaining the status quo cannot be continued due to the huge fiscal deficits of both central and local government in Japan and the severe competitive pressure from other Asian countries.

So it should be necessary for central and local government, universities, and industry to pursue similar policies to the US to create new high-tech industries by the accumulation of university start-up ventures through a new type of networking collaboration, or Triple Helix model, mentioned above, in the regions of Japan. As described earlier, there are quite important segments missing in Japan in the fields of the private equity market and carrier market to make the best and brightest join actively in new venture creation through weak-tie networks. More structural reformation in these fields will be necessary before creating high-tech industries in the Japanese regions. It should be tried, but is quite difficult to predict whether it will be a success or not. There are still challenging tasks ahead for the Japanese economy.

Notes

1 "The sun also rises: a special issue on Japan's economic revival," *The Economist*, October 8–14, 2005, p. 3.
2 Ibid., p. 4.
3 Ibid., p. 6.

4 METI Regional Economy Research Committee *Regional Management under the Declining of Population* (*Jinko genshoka ni okeru chiiki keiei ni tuite* in Japanese), Tokyo: METI, December 2, 2005, pp. 10–15.
5 SME Agency *Japanese SMEs White Paper 2005* (in Japanese), Tokyo: Gyosei, May 20, 2005, pp. 32–5.
6 MITI started to plan a university–industry technology transfer policy based on the report of the survey to compare universities in the US and Japan conducted in 1995 by Arthur D. Little, the US consulting firm, which showed quite impressive results that the Japanese major national universities with huge government support could not make any technological contributions to strengthen the competitiveness of Japanese industry. But this survey missed the close relationship between the individual professor or researcher and the corporation through the Japanese way of collaboration, which was quite informal and seemed to be formalized for creating new collaboration to rejuvenate the Japanese economy based on the research results emerged from the university.
7 Clayton M. Christensen *The Innovator's Dilemma*, Boston: Harvard Business School Press, 1997, pp. 138–41.
8 H. Bremer "History of laws and regulations affecting the transfer of intellectual property", *AUTM Technology Transfer Practice Manual*, Northbrook IL: AUTM, Part 1, 2002, pp. 2–5.
9 H. Etzkowitz *MIT and the Rise of Entrepreneurial Science*, London: Routledge, 2002, p. 16.
10 M. Kenney *Understanding Silicon Valley*, Stanford CA: Stanford University Press, 2000, p. 121.
11 G. Fenn, N. Liang, and S. Prowse *The Economics of the Private Equity Market*, Washington DC: Board of Governors of the Federal Reserve System, 1995, pp. 3–5.
12 W. Bygrave and J. Timmons *Venture Capital at the Crossroad*, Boston: Harvard Business School Press, 1992, pp. 24–5.
13 J. Lerner and F. Hardymon *Venture Capital and Private Equity: A Casebook Volume Two*, New York: Wiley, 2002, p. 160.
14 E. Rogers *Diffusion of Innovations*, 4th edition, New York: Free Press, 1995, pp. 263–4.
15 L. Branscomb and P. Auerswald (eds) *Between Invention and Innovation*, Cambridge MA: US Department of Commerce, 2002, p. 8.
16 H. Etzkowitz, M. Gulbrandsen, and J. Levitt *Public Venture Capital*, 2nd edition, New York: Panel Publishers, 2002, pp. 5–8.
17 P. Eisinger *The Rise of the Entrepreneurial State*, Madison WI: The University of Wisconsin Press, 1988, Chapter 11.
18 This part of the description is based on D. Gibson and E. Rogers *R&D Collaboration on Trial*, Boston: Harvard Business School, 1994, and Larry D. Browning and Judy C. Shetler *Sematech: Saving the U.S. Semiconductor Industry*, College Station TX: Texas A&M University Press, 2000.
19 Gibson and Rogers *R&D Collaboration on Trial*, pp. 453–4.
20 Ibid., pp. 101–2.
21 R. Burt *Structural Holes*, Boston: Harvard University Press, 1992, pp. 45–9.
22 D. Gibson, J. Butler, and T. Kinery "Creating and sustaining the technopolis: Austin, Texas 1985–2003," in A. Nishizawa and M. Fukushima (eds) *University Start-up Ventures and Clustering Strategy* (in Japanese), Tokyo: Gakubunsha, 2005, pp. 56–93.
23 N. Kalis *Technology Commercialization Through New Company Formation: Why US Universities Are Incubating Companies*, Athens OH: NBIA Publications, 2001, Chapter 5.

24 M. Porter "Clusters and the new economics of competition," in J. Magretta (ed.) *Managing in the New Economy*, Japanese version, translated by Diamond Harvard Business Review, Tokyo: Diamond Inc., 2001, p. 76.
25 D. Angel "High-technology agglomeration and the labor market: the case of Silicon Valley", in M. Kenney (ed.) *Understanding Silicon Valley*, Stanford CA: Stanford University Press, 2000, pp. 124–5.
26 M. Granovetter *Getting A Job* (*Tenshoku* in Japanese), Tokyo: Minerva Publishing, 1998, pp. 150–8.
27 S. Tatsuno *The Technopolis Strategy* (*Tekunoporisu Senryaku* in Japanese), Tokyo: Dynamic Sellers, 1988, pp. 86–93.
28 The Japanese central government ministries and agencies were reorganized on January 6, 2001 (www.mofa.go.jp/about/hq/central_gov/index.html).
29 A. Nishizawa "Current situation of venture finance for university spin-off companies in Japan," in D. Gibson, C. Stolp *et al. System and Policies for the Global Learning Economy*, Westport CN: Praeger, 2003, pp. 69–76.
30 M. Castells and P. Hall *Technopoles of the World*, London: Routledge, 1994, pp. 119–26.
31 In 1987, a new regional innovation system was established in Tohoku area called Tohoku Intelligent Cosmos Plan (TICP), which was quite a challenging scheme to create the technological bases for advancement of regional industries through university–industry–government collaboration. But TICP was planned to cover whole region of Tohoku, which was so broad covering seven prefectures and universities locating in these prefectures that any university and local government could not seriously commit to the project without any incentives, and it ceased in early 2006 without any positive results. See S. Abe "Regional innovation systems in Japan: the case of Tohoku", in H. Braczyk, P. Cooke, and M. Heidenreich (eds) *Regional Innovation Systems*, London: UCL Press, 1998, Chapter 12.
32 *Nihon Keizai Shinbun* (*Japan Economic Journal*), January 6, 2006.
33 Nomura Research Institute *Capital Markets Quarterly*, Vol. 9, No. 3, 2005, NRI, Japan, 2005, p. 180.
34 J. Barth, T. Li, S. Malaiyandi *et al.* "Best markets for entrepreneurial finance," *Capital Access Index 2005*, Santa Monica CA: Milken Institute, 2005, p. 3.

8 Cluster classification and performance
The impact of social capital and culture

Bernard Arogyaswamy and
Alojzy Nowak

Introduction

As scholars have stressed[1,2] cities have been integral to the rise of civilizations, an assertion that may apply with equal force to today's "industrial civilization."[3] This civilization cuts across all regional and national boundaries and is a still-evolving outcome of the industrial revolution. Machines (e.g. to produce textiles, to transport goods) operated by steam engines were in the earliest wave of industrialization in Britain, and their manufacture was concentrated around major cities such as London and Manchester. Other cities such as Chicago, Shanghai, Calcutta, Pittsburgh, Milan, and Barcelona (and others too numerous to mention) also grew rapidly in terms of population, outputs, and the accretion of manifold industrial capabilities. Though the days of the sovereign city-state are long gone (apart from possible exceptions such as Singapore and the Vatican), the basic actions—e.g. specialization, labor pools, trade—that led to their success nearly seven centuries ago are no less vital today. The availability of pools of skilled labor, a focus on select products, and trade with other cities were a hallmark of those early cities. Convenient transportation, affordable housing, modern educational institutions, accessible recreational outlets, and so on, characterize today's thriving industrial urban centers. London, Austin, São Paulo, Beijing, and Bangkok exemplify cities that have become centers for production and service-driven businesses.

Industrial civilization

There are, to be sure, ways in which modern cities sharply differ from their historical counterparts, particularly city-states. Given the speed of communications and transportation, skilled people from other parts of the same country and, indeed, from anywhere in the world, can converge on cities such as Austin or Bangalore to take advantage of emerging employment opportunities. Not only is labor mobile, but the technological revolution in information and communication has also given rise to a vastly different "industrial civilization" from that of the manufacturing era. As Friedman[4] notes, technology has made it much easier for developing countries to be partners in the development of the industrial civilization that is taking shape today. Friedman cites a few different ways in which the world is becoming "flatter." For instance, work-flow software makes for seamless linkages among applications in distant locations, open-sourcing provides easy access to software that is being continuously improved, offshoring has given countries with manufacturing/service/technology development capabilities a leg up, while insourcing, on the other hand, enables small firms to leverage the resources of larger corporations in serving remote markets. Though it might seem like a far cry from Braudel to Friedman, both make an almost identical argument: today's civilization is transnational in character. It does not belong to any one nation, region, or religion. It is unifying around a common theme of growth through technology development, sharing, and dispersion. In that sense today's dominant civilization is organized around a second major feature of early civilizations (the first being towns and cities), namely writing, which constitutes, in general, the sharing of knowledge across time (i.e. recording it for future use), and space (with other regions and societies). As we shall see later, the notion of knowledge sharing is a powerful factor in building, and reaping the benefits of, the industrial civilizations of today.

Localization

Paradoxically, the more rapid the pace at which links among globally dispersed operations multiply, and the tighter these bonds become, the more the importance of local capabilities rises, almost in proportion. For one, most nations, realizing the importance of technology to economic growth, would aspire to be at the leading edge of new knowledge or, at the very least, to gain access to it. Equally importantly, transnational corporations (TNCs) are looking for multiple geographic sources of knowledge (based on diverse market experience and local know-how) to achieve sustainable competitive advantages in their enterprise areas. Globalization and localization, therefore, appear to be advancing, so to speak, in lock step. Given the economies of scale in materials and labor, the economies of scope inherent in having access to a range of resources (financing, technology, management, etc.), and lower transaction costs, urban and regional agglomerations are even more attractive today (if one can deal with the problems of overcrowding, pollution, strained infrastructure, and so on) than they were in the past.[5] The "industrial districts" that Marshall[6] wrote about continue to flourish today, and are, in fact, integral to the growth policies pursued by most nations, developed and developing alike. Of course, there are some significant differences. First, today's industrial districts are often consciously developed, frequently with state support and facilitation.[7] Second, due to the diversity and richness of the capabilities agglomerations offer, multiple industrial districts are feasible and, in fact, have been established in many modern cities. London, Beijing, Bangalore, and New York are among the numerous cities with both core and satellite areas of specialization.[8] Ohmae[9] argues that regions (such as Dalian in China) are becoming increasingly important as centers of knowledge and competitive advantage, asserting greater independence from the nations of which they form a part. Their emergence and ties to regions with similar interests suggests intriguing possibilities as relatively confined clusters, the subject of this study, expand geographically.

Business clusters

Perhaps the most common, and arguably most effective, approach to garnering the benefits of localization and physical proximity is in the form of innovative milieux, technology parks, local/regional agglomerations, and so on.[10] There are clear differences among these forms of local/regional groupings of entities that we will attempt to elucidate in the rest of this chapter.

We refer to groupings confined to a relatively small part of a region, and focused on a narrow set of products or services as "business clusters." Typically, business clusters form around a product or technology, with business firms and/or universities, and other institutions, possessing cutting-edge capabilities at their core. Clusters may, however, also be diversified in their activity range, some approximating more closely to the "industrial districts" that Marshall[11] theorized about, while others may spread as they become more successful, to encompass more regional players, or to incorporate more functions vital to the cluster's growth (e.g. greater technological coverage, financial institutions, etc.).[12] Clusters may also grow out of incubators as the firms in the latter succeed and strike out on their own, but within the same locality[13] as in the cases of the Central Florida Research Park and the Delaware Technology Park.[14]

Business clusters are highly variable in form, depending on how they came into being (by design, through evolution, or some combination of both), the nature and intensity of interactions, the types of product(s)/service(s) involved, and so on. While it would be almost impossible to provide an exhaustive typology of clusters, we provide in Figure 8.1 an initial effort at such a cross classification.

Clusters: agglomerations and convergent milieux

Since clusters are theorized to function more effectively in terms of time-to-market, innovativeness, and productivity than stand-alone organizations due to the connections/linkages among the members of the collectivity,[15,16] we employ the number and strength of connections within the cluster as

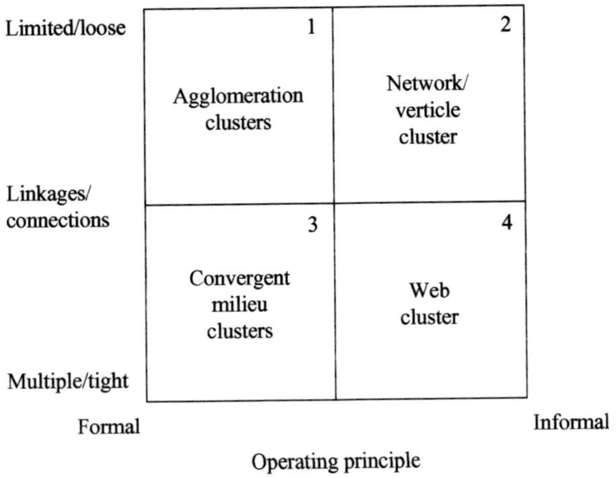

Figure 8.1 Cluster typology
Source: The authors.

one of the dimensions of the typology. Caniels and Romjin[17] categorized clusters into those that are spontaneous and others that are facilitated, offering some guidance in selecting the second dimension. Spontaneity in clusters speaks to whether interactions among the participants occur "passively," that is, whether the collective efficiency (CE) results without any external direction/incentive. In facilitated clusters, on the other hand, Caniels and Romjin define the CE as being active or exogenously coordinated/created. We label the composite of the terms spontaneous/passive and facilitated/active as informal and formal respectively. Where the linkages are loose/limited and the interactions are relatively formal, the cluster or agglomeration (cell 1) is at a rather primitive stage, so to speak. Agglomerations are sometimes found in urban areas that develop an area of specialization. Cities may also, on the other hand, gradually evolve into aggregations in diverse industries. Sometimes this progressive accretion takes place almost at random, resulting in a diverse range of firms being located near one another. Though economies are possible in labor and supplies, these are likely to be rather limited relative to more narrowly focused agglomerations. The relationships within an agglomeration are relatively few and, when they exist, are conducted at arm's length, with little expectation of continued mutual benefit or repeated transactions. In collectivities where the interactions are formal but increasing in number and/or intensity (cell 3), the organizations involved are likely to perceive more mutual benefit from exchanges of materials, ideas, and even employees. Technology parks, often set up or facilitated by governments (or their agencies), are meant to create a regional competence deriving from the proximate existence of educational research, business, financial, and other institutions. We apply the term convergent milieu (CM) to this form of cluster. Milieu is used in quite the same sense as GREMI[18] has attributed to it—a neighborhood or area in which exchanges and transactions are encouraged or have developed spontaneously. However, we do not attach the attribute "innovative" to the milieu as GREMI does, since our rather focused collectivity continues to be formal even if not to the same extent as an agglomeration. "Convergent" obviously refers to the greater commonality of interest among the organizations that have come to occupy locations near one another.

A range of convergent milieux

Manchester's industrial districts, centered around the textile industry in the nineteenth century,[19] Detroit's automobile milieu, and the furniture collective in North Carolina[20] are all examples of an autonomous convergence, that is, one which evolved over a period of time in which the advantages of proximate location became obvious and worked to the mutual benefit of the participants. Most CMs today, however, are designed and implemented by the state or its agencies, Singapore's being early efforts in this direction.[21] China has over fifty science and technology parks, of which many would fall into this category. In particular, we would single out the collectivities in and around Beijing where university–firm interactions are more intensive and frequent than they are in the Shanghai region, which corresponds more to an agglomeration.[22,23] The Chinese parks are designed and organized by the central government (with greater decentralization over time) and include TNCs as prominent players. The relationships, given the relative novelty of the competitive, free market regimen and the fact that most of the technology owners are foreign, are unquestionably formal and lacking in trust.[24] Bangalore, the center of the Indian software industry, may have slightly less formal interactions but it would still be in the CM category. While the main software firms (e.g. Infosys, Wipro, Tata Consultancy) are domestic, the bulk of their customers are located abroad, primarily in the United States, albeit with local offices.[25] The customer (TNC)–supplier (local firm) relationship is likely to become even more formal as the large software firms venture up the "value ladder," going head to head with firms such as Accenture and IBM.[26] Though there is considerable scope for employee mobility among the main providers, the Bangalore software collective is still no more than a rather advanced form of CM, lacking the spontaneous knowledge exchange and "bonding" that take place in close-knit clusters. (Bonding refers to the feeling of community that draws members of a collective to each other in the expectation of mutual benefit based on implicit norms of behavior.[27]) The government of South Korea has, over a period of around two years, helped establish a series of agglomerations and technology centers with a view to creating technological capabilities in industries such as automotives and consumer electronics. In some cases, they have provided incentives for firms to co-locate, for R&D centers to be established at certain places, facilitated the launching of standards associations, etc.[28] The result has been the sprouting of technology parks (CMs) which have fostered the emergence of globally competitive firms in automobiles (e.g. Hyundai) and consumer electronics (Samsung, Gold Star), without any of the clusters becoming informal or trust-based in their transactions.

Beyond convergent milieux

The Swedish pharma aggregation[29] differs from most others in that the Oresmund region (near Denmark) was selected and a brand image care-fully cultivated to attract more firms to an area that already possessed technological capability in the form of leading universities, research insti-tutes, and successful corporations. The telecom milieu in Sweden, on the other hand, developed with the help of a facilitating firm (not focal uni-versity) and through directive corporate leadership. These particular CMs, more than the others, in the area, have progressed to the level of increas-ingly informal, intensive, and mutually beneficial transactions, i.e. beyond the confines of a CM. Similar efforts are underway in other CMs as well, to induce closer relationships so the milieu can develop into a true cluster. The Medtech Collective in Syracuse is an attempt in this direction. The existing CM comprising medical equipment manufacturers, pharmaceu-tical firms, a medical university, and other allied organizations is being configured into a more interactive, even collaborative, milieu by the area's Metropolitan Development Association (MDA). The MDA is also trying to weld the other agglomerations/CMs in the region (metalworking, environmental technology and packaging) into a tighter milieu or cluster.[30]

Clusters: networks and webs

We now shift our attention to a type of regional configuration in which transactions are conducted more informally, with more give and take among the constituents. Networks/vertical clusters (cell 2) involve fewer organizations, which are, however, willing to trust each other to a far greater degree than one would find in agglomerations or CMs.[31] Networks are relatively small-sized collectives (or a subset of a larger grouping) with numerous interconnections among the constituents. The relationships are typically informal and involve low transaction costs and considerable knowledge spillovers.[32] Transaction costs are low because the groundwork has already been laid—supplier's delivery terms and abilities, conditions for bank loans, research institutes procedures, etc., are well known, and do not need to be revisited. Knowledge, particularly tacit knowledge, is shared by design or incidentally, with the expectation that it will work for the good of all. Even where CMs evolve over time indigenously, efforts are often undertaken by external actors (typically state agencies) to induce a more intimate transactional relationship.

When a cluster involves numerous horizontal and vertical linkages, the sharing of valuable, uncodified knowledge assumes greater significance since it calls for a higher level of trust among the protagonists. We label collectives characterized by multiple interactions involving trust in the sharing of knowledge, as "web" clusters in order to emphasize the numerous relationships and implicit understandings the parties have developed with one another. The grouping in Aachen, Germany,[33] and the pharma collective in Oresmund, Sweden[34] have evolved into web clusters. The telecom cluster in Sweden, the Austin computer/information technology cluster, and the North Carolina Technology Park (Research Triangle) are examples of successful actions in this direction.[35]

Trust and social capital

Trust in others' abilities, actions, and assurances is integral to the inter-actions and exchanges that occur in a network or web cluster. Trust resides in the belief that any article or information of value shared with others will not be used to the detriment of the giver. In fact, the expectation is that trust creates a positive-sum game. While trust is the bedrock on which a free market is founded—supported by facilitative, compliance, and punitive mechanisms—it is even more crucial to the functioning of clusters. The balance of competition and cooperation needed for effective knowledge transfer to take place could be irreversibly vitiated if even one serious violation of trust were to occur.

Where vertical relations (e.g. customer–supplier) are involved, contractual trust is essential—both parties must believe that the other will adhere to the terms agreed upon, whether the contract is committed to writing or not. While this may depend, in part, upon the larger society and its ability (e.g. through the judicial system) to enforce contracts, clusters need to evolve their own culture of abiding by the letter and spirit of agreements. Though the steps employed to arrive at greater "contractual trust"[36] might be specific to the context, some of the means could include the creation of standards associations, holding periodic social gatherings, facilitation of the emergence of dominant designs in new products, and so on.

"Competence trust,"[37] which derives from reliance on the other parties' capabilities, is often a matter of prior experiences, testimonials issued, improvements/innovations effected, and so on. Suppliers' success in developing new designs or using better raw materials, creativity of financial institutions in putting together new financing packages, and scientific breakthroughs by universities in the cluster, are all examples of how trust may be established or cemented by demonstrated competence. "Goodwill trust,"[38] unlike the other two varieties, is a pervasive belief that all members in the collective will keep their word, not use information to the detriment of any of them, and will function as though part of a single organic whole.[39,40] Goodwill trust may result from competence and contractual trusts acting, over a period of time, to create an implicit assumption of good faith among the participants in the collective. On the other hand, goodwill trust might be intrinsic to the region's traditional industrial base. In this sense, goodwill trust converges toward the notion of "social capital." We would argue, however, that trust is a *resource* whose deployment creates a *capability* in the form of social capital. As used by authors such as Putnam,[41] and others, social capital speaks to the close ties that develop in certain parts of the world-leading organizations within the region to exchange resources (materials, information, human, managerial) with each other, directly or indirectly, to their mutual benefit. According to Dwivedi, Varman, and Saxena,[42] trust could be based on shared characteristics (family ties, kinship, religion, etc.), common knowledge interests, or interactions with the same institutions. The authors argue that even if trust originated in shared characteristics, its sustainability derives from the parties' track record in knowledge-based transactions or their experiences dealing with common institutions (such as banks, universities, and standards associations). While trust is undoubtedly critical to flexible and fluid exchange relationships, having something valuable to offer to others is equally critical. Knowledge and learning capital[43] are the currency that changes hands in a collective social capital. *Knowledge capital* is the capability to develop new technology in all its facets—product or process, market linkages, arranging for financing, achieving operational efficiencies, and so on—along with the ability and willingness to share relevant technologies with partner organizations.

Knowledge transfer and absorption

Technology developed can be leveraged by sharing, resulting in the creation of new knowledge, or a "surplus," which then may lead to more technology development and so on, in a cycle that parellels resource accumulation in a capitalist system. For investments in the sharing of knowledge to pay off, for the "virtuous cycle" of growth to kick in, however, learning capital is equally indispensable. The ability to learn has to be nurtured as much as the ability to create new technologies. Effective learning, or technology absorption, requires a certain technological convergence (minimal gap in knowledge between giver and receiver) to overcome the not-invented-here (NIH) syndrome, flexibility in internal systems to work with new methods and ideas, decentralized functioning, effective gatekeepers (points of external contact) and boundary spanners (tracking external change), and so on. Knowledge, therefore, is not merely to be "handed off." There are strategies, capabilities, and protocols that need to be adopted by giver and receiver if the transfer is to work to mutual benefit.[44] The store of knowledge and learning capital, and their flow in an atmosphere of trust, determine the reservoir of social capital in any context.

Levels of social capital

Social capital exists at multiple levels—the organizational, regional, national, and global levels. Organizations create capital in numerous ways—by recruiting, training, and retaining competent employees, creating a culture of trust and learning, establishing structures designed to manage knowledge effectively, setting up systems for building and deploying organizational memory, rewarding employees who share their expertise, and so on. Social capital is carefully cultivated and, like technology, is path-dependent.[45] It could be a critical factor in achieving a competitive advantage since organization-specific processes are hard to imitate. While the nurturing of social capital is difficult enough in an organization at a specific location, the problems multiply with distance. However, through the use of information technology, incentives, flexible structures, etc., firms such as Accenture and Nucor have attempted to foster knowledge transfer and learning across their far-flung operations.[46,47] While the friction, so to speak, in the flow of knowledge within organizations may be overcome through various intrinsic (e.g. by fostering a sense of belonging) and extrinsic (e.g. incentives) means, creating and sustaining inter-organizational social capital, on a limited or extensive scale, is a quite different proposition. First it requires that organizations that participate in the relationship need to have *internal* social capital on which they may build external ties. Unless employees and departments trust each other enough to trade knowledge without always themselves benefiting from such a transfer, and are capable of using the knowledge received, they are unlikely to exhibit the same behavior externally. For external social capital to be created, therefore, trust and knowledge exchange capabilities within organizations must be actively cultivated.

Organizational and societal cultures

There is, however, an equally important factor at play here—the culture of the society to which the cluster belongs. Hofstede,[48] Trompenaars, and Hampden-Turner,[49] and others have provided us with dimensions of culture that have emerged from their empirical work. Among the distinguishing aspects of a culture they identify are the degrees of individualism, PD, universalism, uncertainty avoidance (UA), and so on, which characterize a particular society. If, for instance, the national culture emphasized individualism (in terms of effort, achievement, and reward), high PD, and an adherence to written rules and procedures (one aspect of UA), one might anticipate a reluctance to trust others in the society at large. Schneider and Barsoux[50] posit the existence of four types of societal cultures based on varying levels of PD and UA. While the types of cultures are labeled illustratively (those low on PD, high on UA are called "well-oiled machines," while those high on PD, low on UA are termed the "families/tribes"), the inescapable conclusion is that cultures have strong distinguishing characteristics, which may influence the functioning of organizations and, by extension, clusters. While it might be possible to create a CM (say, in the form of technology parks), moving it along to the network or web stage of clustering might prove to be an uphill task. For instance, cultures that are high on both PD and UA might find it particularly difficult to develop clusters marked by large reserves of social capital—the hierarchic emphasis combined with rule orientation and risk aversion might require special efforts to overcome the "cultural inertia" inhibiting trust formation and the sharing of knowledge.

Occupational cultures

Firms' modes of functioning, as emphasized above, significantly impact the store and flow of social capital in networks and webs. Corporate cultures, to some degree, are creatures of the larger societal culture within which they function. For instance, societal cultures characterized by low trust are likely to be home to organizations reflecting this trait. It must be made clear, however, that corporations need not always faithfully reflect societal culture. Occupational culture could be a powerful mediating variable.[51] In industries in which the core functions are predictable and programmed, hierarchic and standardized methods of working are likely to predominate. Specialized departmentation could result in barriers to horizontal communication, lack of vertical and horizontal trust, and so on. "Traditional" manufacturing industries (e.g. mass production—automobiles, appliances, etc.) would fall into this category though there may well be attempts made to alter the culture to stimulate employee involvement and decision-making, encourage freer circulation of ideas and information, and overall, induce an organic mode of functioning. The presence of many, or even a few, "trust-inhibited" organizations could slow down or even prevent the formation of social capital in any milieu. The occupational culture of knowledge-intensive industries, however, is typically likely to be far more adaptive and flexible to markets and employee needs, providing a context in which decisions are made based on the competence of the decision-maker not the level or position), and where horizontal and vertical differentiation is minimized through a climate and structure of collaboration. While economies as a whole are undoubtedly becoming increasingly knowledge-intensive (knowledge input constitutes over 70 percent of the US economy, as opposed to 40 percent a bare quarter century ago),[52] industries such as information technology, biotechnology, nanotechnology, and aerospace require ever high levels of technology creation and absorption, making a constant search for, and adaptation to, new ideas imperative.[53] The cultures of firms in these industries, therefore, are likely to be far more oriented to knowledge creation and a greater level of comfort with the interdependencies that are integral to achieving this end.

Developing cultures supportive of social capital

Certain societies, which possess a strong community feeling, have successfully parlayed their social capital into a competitive advantage. On the other hand, in Silicon Valley, an area with no preexisting tendencies to bonding, trust and social capital evolved, in part, due to the technological (and cultural) sway held by innovative institutions such as Stanford University, Hewlett Packard, Fairchild Semiconductor, and others that created a supportive, sharing culture in the area.[54] Of course "confidence building" mechanisms such as industry associations, club memberships, labor mobility, and so on, helped, but the cluster culture was determined by the larger firms, and the smaller ones were acculturated, so to speak, into the system.[55] While industries such as tile making, footwear, and furniture are no doubt becoming increasingly knowledge-driven, there is an order of magnitude difference separating them from, say, information technology and biotechnology, in terms of knowledge content and its rate of change. "Knowledge spillovers" are often cited as a hallmark of functioning clusters (though the term seems to connote, at least in part, an accidental and involuntary transfer). In high-tech areas, knowledge transfers may be characterized by economies of scale (frequent interactions within a network) and/or of scope (across a wider web of organizations), but the more voluntary the sharing, the more likely that tacit knowledge will flow within the cluster. It may be noted that tacit knowledge applies equally to "hard" areas such as product/process technology and to "soft" areas such as market research, employee motivation, promotion, and so on. The paradox is that as knowledge capital becomes more technically specialized it magnifies the need for a feeling of trust among the members of a collective. "High tech, high touch"[56] is indeed a reality. The trust does not, of course, arise out of altruism. Few individuals and even fewer organizations behave completely selflessly. The instances of altruism in nature, for instance, are predicated upon raising the survival chances of the species (or a subset of it), due to which the individual animal might be willing even to sacrifice itself.[57] Though drastic actions of this nature are hardly called for in the ecology of a cluster, the principle of mutual preservation, and group success through reciprocal altruism hold ("You scratch my back, I scratch your back").[58,59]

"Brownfield" clusters

More collectives are, to be sure, brought into existence by entities such as the state, rather than arising and evolving autonomously. Some are created from scratch (e.g. the North Carolina Research Triangle, the numerous science and technology parks in China, Sweden's aluminum, telecom, and other clusters, etc.) while others, already in existence, are sought to be brought together by instilling a sense of internal bonding. The Medtech grouping in Syracuse, New York, cited earlier, provides an instance of the latter. The local MDA is attempting to weld the already existing CM comprising around sixty-plus medical device/appliance companies into a web cluster. A coordinator has been appointed charged with organizing events to familiarize members with one another, serving as a clearinghouse of ideas, organizing seminars on matters of mutual technological and market interest, enhancing the attractiveness of the region to highly qualified people, and so on.[60]

The problem with creating a culture in which there is sufficient social capital is common to both the "greenfield" (from the ground up) and "brownfield" (in an existing grouping), so to speak, milieux. In fact, the brownfield version (binding an existing CM together, as in Medtech) probably poses an even bigger hurdle since the participants have likely operated in a climate of indifference or even hostility. Be that as it may, changing the culture of a collective to one supportive of a network or web is at least as difficult as changing a corporate culture when the firm's strategy changes.[61,62] While, in an organizational context, culture change may be driven by executive communication and example, incentives, team-based exercises, and so on, the disparate members of a collective are not likely to be easily swayed by a corporate leader, external coordinator, or through the medium of common events, associations, and issues.

Initial attempts should, logically, be on a rather small scale. For instance, given that larger firms (Welch Allyn or Sensus in Medtech) invest considerable sums on R&D and fear the leakage of proprietary knowledge, their ties to smaller firms need to go through a series of confidence-building measures, to establish both competence and contractual trust. While a "coordinator" is very useful for the proposed cluster, efforts ought to be directed to creating this sort of trust, two firms at a time. Other activities such as organizing seminars and discussions on a broader scale can, however, proceed in parallel. The onus is typically on the larger firm, both to establish a two-way partnership as well as to induce a climate of trust that might attract other entrepreneurs to the vicinity. Convincing skeptical leaders of the bigger corporations of the wisdom of sharing not just technical knowledge but also how best to manage employees, arrange financing, cooperate with universities and research institutes, etc., is one of the critical tasks of culture-building in a cluster. Subsequently, expanding the reach of interactions beyond the network stage to include horizontal connections, including those among larger firms and among their smaller partners as well, would be a second phase in the effort to make an organic web of relationships a reality.

Cluster performance

In developing our taxonomy of clusters, we have employed dimensions such as range of activities, geographic spread, nature of connections, and the operating principle (formal vs. informal) to help distinguish among the numerous and varied agglomerations worldwide. While these factors are no doubt useful in helping to differentiate among clusters and to get a better sense of their internal processes, they do not give us a sense of their effectiveness. That is, though there might be an appropriate focus, intense interactions, and informal relationships within a cluster, its ability to satisfy the needs of its constituents and to achieve the performance desired (as in Figure 8.2) might well be in doubt. Among the strategic goals of any cluster, we would include:

1   Growth in membership; low turnover.
2   Innovation in terms of patents received and R&D expenditures (say, per 1,000 employees).
3   Value added within the cluster.
4   Employment within the cluster; also employment generated in surrounding areas.
5   Total sales achieved to external customers.

Obviously, context and stage of evolution will determine which goals are most appropriate. For instance, in the early phases soon after a cluster is established, membership and low turnover would be far more appropriate indicators than at maturity. Innovation measures would be more relevant in high-tech clusters while value added within would be important when large buyers are located in the cluster itself.

Various propositions may be developed relating cluster performance to the contextual factors discussed earlier. A few are provided here for illustrative purposes:

Proposition 1
Clusters in which interactions among organizations are spontaneous, and which involve the exchange of tacit knowledge, will perform better

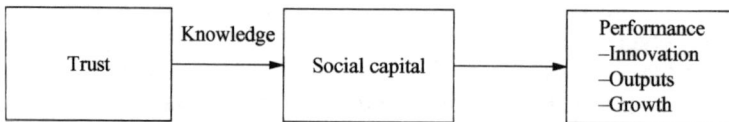

Figure 8.2 Cluster performance

Source: The authors.

(in terms of growth and external sales) than clusters requiring exogenous coordination and involving little exchange of knowledge.

Proposition 2
Clusters in which large organizations are low in social capital (e.g. centralized) are likely to be low performing clusters.

Proposition 3
In cultures that are high on both PD and UA, clusters are likely to be low in performance.

Proposition 4
More frequent and spontaneous exchange of tacit knowledge will be more strongly associated with high cluster performance in high technology areas than for low technology businesses.

Institutions and social capital in clusters

While business firms are, no doubt, an important part of the culture and social capital that characterize a cluster, other institutions are no less vital. Educational, financial, technological, and political institutions come to mind in this context. All societies are supported by a set of institutions, which are, in the main, learnt patterns of behavior and thought, and which help deal with issues that arise in everyday life.[63] Whether in banking, religion, or education, institutions embody past standard practice, and tend to resist change. Institutional economists such as Hamilton[64] and Hollingsworth[65] have argued that a range of societal institutions along with the traditions and practices that have brought countries to where they are, may need to be changed if progress is to be achieved. Veblen[66] had viewed economics as an evolutionary process in which change was gradual, technology being the potential for change, and institutions comprising the static element to which technology typically had to adapt itself. While Hamilton's perspective was more optimistic, suggesting that technology could act upon and change societal institutions, nevertheless, for economic progress to be achieved through the means of technology, independent institutional change is an essential prerequisite and concomitant. Martin, Velasquez, and Funck[67] implicitly echo this view in their report titled, "Building institutions for markets," in which present and future efforts to initiate institutional change in diverse, mainly developing, nations are detailed.

In the evolution of clusters, universities, research institutions, banks, venture capitalists, consultants, and labor unions are among the institutions that can play a vital part.[68] Since these institutions are a subset of the larger body of national institutions, they are likely to reflect the national "ethos" or "mind-set".[69] It is quite possible that in reflecting the national mind-set, such institutions could accelerate or derail the incremental trust-creation that must take place in a potential cluster. For instance, research institutes in Poland, in part due to Communist-era policies, have typically not involved themselves in the marketization of products but have confined themselves to the technology development phase. Collaboration with business firms has been minimal, the ties between business and universities being even weaker. In India, on the other hand, research institutes (such as the Indian Institute of Science) and universities have become increasingly engaged with the private sector particularly in high-tech fields such as information technology and biotechnology. However, banks still retain vestiges of having been nationalized, and hence are less flexible and adaptive to the needs of cluster development than their counterparts in many other countries. Depending on the broader, national culture, the type and degree of institutional change needed (within a cluster) will obviously vary. For instance, in the family/tribe culture,[70] webs might arise autonomously, provided they remain confined to the traditional or "primeval"[71] participants, while in family groupings, networks could function effectively as in the Chinese diaspora in parts of Southeast Asia.

Conclusion

Globalization is a force that has exercised powerful sway over our lives. Though its impacts on lifestyles, labor, the environment, prosperity, disparity, and other issues continue to be debated, there is little doubt that the phenomenon of globalization is here to stay. There is also little doubt that the driving force behind globalization is technology, which is leading to the emergence of a common industrial civilization. While TNCs are often viewed as the engines of "techno-globalization," so to speak, nation-states are no less active in attempting to create knowledge capabilities in a variety of industries within their shores. Urban and regional collectives are a popular approach to demonstrating such a capability, both to bring existing talent together as well as to attract competent individuals and organizations from outside the area. The types of organizations needed, the role of the state, and the importance of trust and social capital have been extensively discussed by other authors. The contribution of this chapter lies, first, in the classifications developed of the varieties of clusters that may evolve or be constructed by design, and second, in developing illustrative measures of cluster performance. A third strand of reasoning articulated here is that social capital, deemed critical to the successful operation of clusters, is conditioned upon cultures at various levels. While the three concepts are, no doubt, linked, each may be profitably pursued in other inquiries, both theoretical and empirical. The notion of institutional reform within clusters is proposed as a mechanism by which culture change may be effected locally even if the larger culture (of the nation, of firms based therein, or of TNCs with a unified culture) militates against the creation of social capital.

Notes

1 D. S. Landes *The Wealth and Poverty of Nations*, New York: W. W. Norton, 1998.
2 P. Jay *The Wealth of Man*, New York: Public Affairs, 2000.
3 F. Braudel *A History of Civilizations*, New York: Penguin, 1993.
4 T. L. Friedman *The World is Flat: A Brief History of the Twenty-first Century*, New York: Fairar, Strauss and Giroux, 2005.
5 A. Cambers and D. Mackinnon "Clusters in urban and regional development," *Urban Studies*, Vol. 41, Nos. 5/6, May 2004, pp. 959–69.
6 A. Marshall *Industry and Trade*, 4th edition, London: Macmillan & Co., 1923.
7 M. Porter *The Competitive Advantage of Nations*, Basingstoke: Macmillan, 1990.
8 P. Cooke, C. Davies, and R. Wilson "Innovation advantages of cities: from knowledge to equity in five basic steps," *European Planning Studies*, Vol. 10, No. 2, 2002, pp. 233–50.
9 K. Ohmae *The Next Global Stage: Challenges and Opportunities in Our Borderless World*, Singapore: Pearson Education, 2005.
10 M. Perry *Business Clusters: An International Perspective*, London: Routledge, 2005.
11 Marshall *Industry and Trade*, 1923.
12 S. A. Rosenfeld "Expanding opportunities: cluster strategies that reach more people in more places," *European Planning Studies*, Vol. II, No. 4, 2003, pp. 359–77.
13 CFRP, 2005, www.cfrp.org.
14 Deltechpark, 2005, www.deltechpark.org.
15 P. Cooke *Knowledge Economies: Clusters, Learning and Cooperative Advantage*, London: Routledge, 2002.
16 M. Porter "Clusters and the new economies of competition," *Harvard Business Review*, Vol. 76, November/December 1998, pp. 77–90.
17 M. C. Caniels and H. A. Romjin "Dynamic clusters in developing countries," *Oxford Development Studies*, Vol. 31, No. 3, September 2003, pp. 275–92.
18 Perry *Business Clusters*, 2005.
19 Ibid.
20 National Governors Association *A Governor's Guide to Cluster-based Economic Development*, 2002, www.nga.org.
21 D. Mackendrick, R. Doner, and S. Haggard *From Silicon Valley to Singapore*, Stanford CA: Stanford University Press, 2000.
22 S. M. Walcott *Chinese Science and Technology Industrial Parks*, Aldershot: Ashgate Publishing, 2003.
23 D. Waldron "Growth triangles: a strategic assessment," *Multinational Business Review*, Vol. 5, No. 1, Spring 1997, pp. 53–67.
24 Walcott *Chinese Science and Technology Industrial Parks*, 2003.
25 Caniels and Romjin "Dynamic clusters in developing countries," 2003.
26 Business Week "The future of outsourcing", 2006, www.businessweek.com/magazine/content/06_05/b3969401.htm.
27 M. Storper "Society, community, and economic development," *Studies in Comparative Economic Development*, Vol. 39, No. 4, Winter 2005, pp. 30–57.

28 K. Lee "Promoting innovative clusters through the Regional Research Center (RRC) policy programme in Korea," *European Planning Studies*, Vol. 11, No. 1, 2003, pp. 25–39.
29 P. Lundequist and D. Power "Putting Porter into practice? Practices of regional cluster building: evidence from Sweden," *European Planning Studies*, Vol. 10, No. 6, 2002, pp. 685–704.
30 MDA *The Essential New York Initiative*, February 26, 2004, www.cnymda.org.
31 Cooke *Knowledge Economies*, 2002.
32 Cambers and Mackinnon "Clusters in urban and regional development," 2004.
33 R. Sternberg and T. Litzenberger "Regional clusters in Germany—their geography and their relevance for entrepreneurial activities," *European Planning Studies*, Vol. 12, No. 6, 2004, pp. 767–91.
34 Lundequist and Power "Putting Porter into practice?," 2002.
35 Perry *Business Clusters*, 2005.
36 M. Sako *Prices, Quality and Trust*, Cambridge: Cambridge University Press, 1992.
37 Ibid.
38 Ibid.
39 N. Lukmann *Trust and Power*, Chichester: Wiley, 1979.
40 C. Sabel "Studied trust: building new forms of cooperation in a volatile economy," in F. Pyke and W. Sengenberger (eds) *Industrial Districts and Local Economic Regeneration*, Geneva: IILS, 1992, pp. 215–50.
41 R. Putnam *Making Democracy Work: Civic Traditions in Modern Italy*, Princeton NJ: Princeton University Press, 1993.
42 M. Dwivedi, R. Varman, and K. Saxena "Nature of trust in small firm clusters," *The International Journal of Organizational Analysis*, Vol. II, No. 2, 2003, pp. 93–104.
43 Cooke *Knowledge Economies*, 2002.
44 B. Arogyaswamy and W. Elmer "Technology absorption in emerging nations: an institutional approach," *Journal of East-West Business*, Vol. 10, No. 4, 2004, pp. 79–104.
45 Storper "Society, community, and economic development," 2005.
46 M. T. Hansen, N. Nohria, and T. Tierney "What's your strategy for managing knowledge?," *Harvard Business Review*, Vol. 77, No. 2, 1999, pp. 106–18.
47 A. K. Gupta and V. Govindarajan "Knowledge management's social dimension: lessons from Nucor Steel," *Organizational Dynamics*, Fall 2000, pp. 71–80.
48 G. Hofstede *Culture's Consequences: International Differences in Work-related Values*, Beverly Hills CA: Sage, 1980.
49 F. Trompenaars and C. Hampden-Turner *Riding the Waves of Culture*, New York: McGraw-Hill, 1998.
50 S. Schneider and J. L. Barsoux "Culture and organization," in S. Schneider and J. L. Barsoux (eds) *Managing Across Cultures*, London: Prentice-Hall, 1997, pp. 87–95.
51 S. S. Cohen and G. Fields "Social capital and capital gains: an examination of social capital in Silicon Valley," in M. Kenney (ed.) *Understanding Silicon Valley*, Stanford CA: Stanford University Press, 2000, pp. 190–217.
52 Cooke, Davies, and Wilson "Innovation advantages of cities," 2002.
53 P. Cooke "Regional knowledge capabilities, embeddedness of firms, and industry organization: bioscience megacenters and economic geography," *European Planning Studies*, Vol. 12, No. 5, 2004, pp. 625–41.
54 T. Sturgeon "How Silicon Valley came to be," in M. Kenney (ed.) *Understanding Silicon Valley: The Anatomy of an Entrepreneurial Region*, Stanford, CA: Stanford University Press, 2000, pp. 15–47.

55 C. B. Schoonoven and K. Eisenhardt "The impact of incubator region on the creation and survival of new semiconductor ventures in the U.S. 1978–1986," Report to the EDA, US Department of Commerce (August 1989).
56 J. Naisbitt *Megatrends*, New York: Warner, 1982.
57 J. Casti *Paradigms Lost*, New York: Avon, 1989.
58 Cooke *Knowledge Economies*, 2002.
59 Casti *Paradigms Lost*, 1989.
60 CNY Medtech, 2006, www.cnymedtech.org.
61 B. Arogyaswamy and C. Byles "Organizational culture: internal and external fits," *Journal of Management*, Vol. 13, No. 4, Winter 1987, pp. 647–58.
62 C. Byles, K. Aupperle, and B. Arogyaswamy "Organizational culture and performance," *Journal of Managerial Issues*, Vol. 3, No. 4, Winter 1991, pp. 512–27.
63 B. Loasby *Knowledge, Institutions, and Evolution in Economics*, London: Routledge, 1999.
64 D. Hamilton *Evolutionary Economics: A Study of Change in Economic Thought*, Albuquerque NM: University of New Mexico Press, 1970.
65 R. Hollingsworth *Doing Institutional Analysis: Implications for the Study of Innovation*, mimeo, Madison WI: University of Wisconsin, 1998.
66 T. Veblen *The Instinct of Workmanship*, reprint, New York: W. W. Norton and Co., 1964.
67 C. Martin, F. Velasquez, and B. Funck *European Integration and Income Convergence: Lessons for Central and Eastern European Countries*, Washington DC: The World Bank, 2001.
68 M. Kenney and U. VonBurg "Institutions and economies," in M. Kenney (ed.) *Understanding Silicon Valley*, Stanford CA: Stanford University Press, 2000, pp. 218–40.
69 J. Stiglitz "Participation and development: perspectives from the comprehensive development paradigm," in F. Iqbal and J. Yon (eds) *Democracy, Market Economies, and Development: An Asian Perspective*, Washington DC: The World Bank, 2001, pp. 49–72.
70 Schneider and Barsoux "Culture and organization," 1997.
71 Storper "Society, community, and economic development," 2005.

Index

Pages containing relevant illustrations are indicated in *italic* type.